Springer Tracts in Modern Physics
Volume 203

Springer Tracts in Modern Physics

Springer Tracts in Modern Physics provides comprehensive and critical reviews of topics of current interest in physics. The following fields are emphasized: elementary particle physics, solid-state physics, complex systems, and fundamental astrophysics.
Suitable reviews of other fields can also be accepted. The editors encourage prospective authors to correspond with them in advance of submitting an article. For reviews of topics belonging to the above mentioned fields, they should address the responsible editor, otherwise the managing editor.
See also springeronline.com

Thomas Mannel

Effective Field Theories in Flavour Physics

With 29 Figures

Professor Thomas Mannel
Theoretische Physik 1
Emmy Noether Campus
Universität Siegen
57068 Siegen, Germany
E-mail:mannel@hep.physik.uni-siegen.de

Library of Congress Control Number: 2004106871

Physics and Astronomy Classification Scheme (PACS):
11.10.Ef, 11.30.Hv, 12.39 Hg

ISSN print edition: 0081-3869
ISSN electronic edition: 1615-0430
ISBN 3-540-21931-5 Springer Berlin Heidelberg New York

Springer is a part of Springer Science+Business Media

springeronline.com

Typesetting: by the author and TechBooks using a Springer LaTeX macro package
Cover concept: eStudio Calamar Steinen
Cover production: *design &production* GmbH, Heidelberg

Printed on acid-free paper SPIN: 10572279 56/3141/jl 5 4 3 2 1 0

Preface

This book emerged from a long process of trying to write a monograph on the experimental and theoretical aspects of flavour physics, with some focus on heavy-flavour physics. This original scope turned out to be far too wide and had to be narrowed down in order to end up with a monograph of reasonable size. In addition, the field of flavour physics is evolving rapidly, theoretically as well as experimentally, and in view of this it is impossible to cover all the interesting subjects in an up-to-date fashion.

Thus the present book focuses on theoretical methods, restricting the possible applications to a small set of examples. In fact, the theoretical machinery used in flavour physics can be summarized under the heading of effective field theory, and some of the effective theories used (such as chiral perturbation theory, heavy-quark effective theory and the heavy-mass expansion) are in a very mature state, while other, more recent ideas (such as soft-collinear effective theory) are currently under investigation.

The book tries to give a survey of the methods of effective field theory in flavour physics, trying to keep a balance between textbook material and topics of current research. It should be useful for advanced students who want to get into active research in the field. It requires as a prerequisite some knowledge about basic quantum field theory and the principles of the Standard Model.

Many of my colleagues and students have contributed to the book in one way or another. In the early stages, when the scope was still defined very widely, I enjoyed discussions with Ahmed Ali and Henning Schröder. In the later stages I had some help from Wolfgang Kilian, Jürgen Reuter, Alexander Khodjamirian, Heike Boos and Martin Melcher, some of who were "test persons", who told me, which parts of the book were still incomprehensible.

Finally, I want to thank my wife Doris and my children Thurid, Birte and Hendrik for their patience; a lot of time, which should have been dedicated to them, went into writing this book.

Siegen, September 2004 *Thomas Mannel*

Contents

1 Introduction

1.1 Historical Remarks

The beginning of flavour physics can be dated back to the discovery of nuclear β decay by Becquerel and Rutherford in the late nineteenth century [1, 2]. Almost twenty years later it was noticed by Chadwick [3] that the "β rays" had a continuous energy spectrum, which was at that time a complete mystery. The measurements of the nuclear β decay

$$^{210}_{83}\,\mathrm{Bi} \longrightarrow {}^{210}_{84}\,\mathrm{Po}$$

by Ellis and Wooster in 1927 [4] showed an average electron energy $\langle E_\beta \rangle = 350\,\mathrm{keV}$, while the mass difference of the two nuclei is $E_\beta^{\mathrm{max}} = 1050\,\mathrm{keV}$. This result was indeed mysterious, since it would imply the violation of energy conservation.

In order to save energy conservation, Pauli postulated the existence of a particle that escaped observation. In his famous letter to the "Radioaktive Damen und Herren" (a reprint of this letter can be found in [5]) he postulated the neutrino, the interactions of which had to be so weak that it did not leave any trace in the experiments which could be performed at that time. It took more than twenty years to find direct evidence for the neutrino: in 1953 the process $\bar{\nu}_e + p \to n + e^+$ was observed by Reines and collaborators [6, 7, 8].

On the theoretical side, the description of weak interactions started in 1933 with Fermi's idea of writing the interaction for β decay as a current–current coupling [9]. Motivated by the structure of electrodynamics, he wrote the interaction for the β decay of a neutron as

$$H_{\mathrm{int}} = G \int d^3x \, [\bar{p}(x)\gamma_\mu n(x)][\bar{e}(x)\gamma_\mu \nu(x)] \, , \tag{1.1}$$

since at that time the proton and the neutron were considered as elementary spin-1/2 particles. Comparison with data at that time showed that the neutrino mass was small compared with the electron mass and that the value of the Fermi coupling was $G \approx 0.3 \times 10^{-5}\,\mathrm{GeV}^{-2}$.

With more precise data on nuclear β decays, inconsistencies with the simple ansatz (1.1) became apparent and a generalization was necessary. Gamov suggested in 1936 [10] that (1.1) should be generalized to

$$H_{\text{int}} = \sum_j \int d^3x\, g_j [\bar{p}(x) M_j n(x)][\bar{e}(x) M_j \nu(x)] \,, \tag{1.2}$$

where the M_j run over the set of Dirac matrices

$$M_j \otimes M_j = 1 \otimes 1 \,,\ \gamma_5 \otimes \gamma_5 \,,\ \gamma_\mu \otimes \gamma^\mu \,,\ \gamma_\mu\gamma_5 \otimes \gamma^\mu\gamma_5 \,,\ \sigma_{\mu\nu} \otimes \sigma^{\mu\nu} \,, \tag{1.3}$$

and the g_j are real parameters. The choice of (1.3) assumes that the discrete symmetries parity (P) and charge conjugation (C) hold separately, which was a standard assumption at that time, since the observed strong and electromagnetic interactions conserved those symmetries. Under these assumptions H_{int} in (1.2) is the most general ansatz.

It was also noticed quite early that the strengths of the weak processes known at that time were very similar. After the discovery of the pion and the muon, the couplings of $n \to pe\bar{\nu}_e$, $\pi \to \mu\bar{\nu}$ and $\mu \to e\bar{\nu}\nu$ turned out to be similar, once an ansatz similar to (1.1) or (1.2) was taken for the pion and muon decays [11]. This was taken very early on as a hint that weak interactions were governed by some kind of universality. However, in the 1950's it became clear that nuclear β decay was well described by (1.2) using a combination of $1 \otimes 1$ and $\sigma_{\mu\nu} \otimes \sigma^{\mu\nu}$, while muon decay was best described by a combination of $\gamma_\mu \otimes \gamma^\mu$ and $\gamma_\mu\gamma_5 \otimes \gamma^\mu\gamma_5$. Consequently, the universality of weak interactions became questionable.

At about the same time, new particles were observed which showed a strange behaviour. Being heavier than three pions they were expected to decay strongly into two or three pions. These were indeed the main decay modes of these particles, however, but they had a lifetime typical of a weak process. These particles also triggered another breakthrough in our understanding of weak interactions, which was called the $\Theta - -\tau$ puzzle. The Θ and τ were particles with decay modes $\Theta \to \pi^+\pi^0$ and $\tau \to \pi^+\pi^+\pi^-$, which means that their final states have different parities, assuming an s-wave decay. The puzzle consisted in the fact that the Θ and τ had the same mass and lifetime within the accuracy of the measurements, but different parities.

The solution of this puzzle was given by Lee and Yang in 1956 [12], who postulated that the Θ and τ are identical; in today's naming scheme, this particle is the K^+. This implied the bold assumption that weak interactions violate parity, which was considered unacceptable by many colleagues at the time. However, soon after the idea of Lee and Yang, parity violation was experimentally verified by Wu et al. [13] and Garwin et al. [14] in 1957.

For the theoretical description this means that (1.2) has to be modified again to accommodate parity violation. The best fit for neutron β decay is obtained with

$$H_{int} = -\frac{G_\beta}{\sqrt{2}} \int d^3x \left[\bar{p}(x)\gamma_\mu \left(1 - \frac{g_A}{g_V}\gamma_5\right) n(x)\right] [\bar{e}(x)\gamma^\mu(1-\gamma_5)\nu(x)] \,, \tag{1.4}$$

which is basically today's description of neutron β decay; the values of the parameters are

$$
\begin{aligned}
G_\beta &= (1.14730 \pm 0.00064) \times 10^{-5}\ \mathrm{GeV}^{-2}\ , \\
\frac{g_A}{g_V} &= 1.255 \pm 0.006\ .
\end{aligned}
\tag{1.5}
$$

From today's point of view, the fact that $g_A/g_V \neq 1$ comes from the fact that neither the proton nor the neutron is an elementary particle.

After implementing parity violation, it became clear that pion, muon and neutron weak decays are basically described by a "vector minus axial vector" $(V - A)$ current–current coupling with the same coupling constant for all these decays. Weak interactions again exhibited universality.

The next breakthrough in weak-interaction physics came again from the strange particles mentioned above. In the 1950's the "particle zoo" developed, staring with the kaons and other strange particles. The lifetimes of these particles turned out to be long compared with typical lifetimes for strongly decaying states, so decays such as $K^+ \to \pi^+\pi^0$ were identified with weak decays. This was implemented by postulating a new quantum number S ("strangeness") [15, 16, 17], which is conserved in strong processes but may change in weak processes.

From weak-interaction universality, one would conclude that the strangeness-changing processes should have the same coupling strength as the strangeness-conserving ones, for example the coupling for $K^+ \to \pi^+\pi^0$ should be the same as for $\pi \to \mu\bar{\nu}$. This turned out to be grossly wrong: the rates for strangeness-changing processes are suppressed by about a factor of 20 compared with the strangeness-conserving ones. This contradicted the concept of universality of weak interactions.

In 1963, universality was resurrected by Cabibbo [18], who used current algebra to argue that the total hadronic $V - A$ current H_μ should have "unit length", i.e.

$$
H_\mu = H_\mu^{\Delta S=0} \cos\Theta + H_\mu^{\Delta S=1} \sin\Theta\ , \tag{1.6}
$$

where $H_\mu^{\Delta S=0}$ is the hadronic current for strangeness-conserving processes, $H_\mu^{\Delta S=1}$ governs the strangeness-changing decays and Θ is the Cabibbo angle. Experimentally it was found that $\sin\Theta \approx 0.22$, which explained all strangeness-changing processes consistently. Up to the rotation (1.6), weak interactions were again universal.

A further step in developing our present understanding was the discussion of neutral currents. Up to that point the weak $V-A$ currents were all charged currents, i.e. they connected particles which differed by one unit of charge. Generically one would also expect neutral currents of similar strength, in particular flavour-changing neutral currents. However, it was noticed quite early on that

$$
\frac{\Gamma(K^+ \to \pi^+\nu\bar{\nu})}{\Gamma(K^+ \to \pi^0 e^+\bar{\nu})} < 10^{-5} \ll 1\ , \tag{1.7}
$$

which implied a strong suppression of flavour-changing neutral processes.

The large zoo of particles was ordered once the quark substructure of hadrons had been noticed [19]. Although at first it was only a model used for the classification of the hadronic states, it also put the weak interactions into a different perspective. Weak processes were understood as transitions between different quark flavours. The hadronic current in (1.6) is written in modern language as

$$H_\mu = \bar{u}\gamma_\mu(1-\gamma_5)\left[d\cos\Theta + s\sin\Theta\right] \ , \tag{1.8}$$

where the s quark carries the strangeness quantum number -1. Consequently, it is the combination $[d\cos\Theta + s\sin\Theta]$ which participates in the weak interaction.

Given this point of view, a neutral current would have the general form

$$H_{\text{neutral}} = \bar{u}Mu + \left[\bar{d}\cos\Theta + \bar{s}\sin\Theta\right] M' \left[d\cos\Theta + s\sin\Theta\right] \ , \tag{1.9}$$

where M and M' are some Dirac matrices. Clearly (1.9) exhibits a $\Delta S = \pm 1$ contribution

$$H_{\text{neutral}}^{\Delta S=\pm 1} = \cos\Theta\sin\Theta\left[\bar{d}M's + \bar{s}M'd\right] \ , \tag{1.10}$$

which would make the ratio (1.7) of order unity, in contradiction with observations.

The solution to this problem was found by Glashow, Ilipoulos and Maiani [20] and is called the GIM mechanism. Their idea had the surprising consequence that another quark with the quantum numbers of the up quark has to exist. This new quark, the charm quark, couples to the "orthogonal" combination $[s\cos\Theta - d\sin\Theta]$ of the down and the strange quark. While the charged current becomes

$$H_\mu = \bar{u}\gamma_\mu(1-\gamma_5)\left[d\cos\Theta + s\sin\Theta\right] + \bar{c}\gamma_\mu(1-\gamma_5)\left[s\cos\Theta - d\sin\Theta\right] \ , \tag{1.11}$$

the neutral current is

$$\begin{aligned} H_{\text{neutral}} &= \bar{u}Mu + \bar{c}Mc \\ &\quad + \left[\bar{d}\cos\Theta + \bar{s}\sin\Theta\right] M' \left[d\cos\Theta + s\sin\Theta\right] \\ &\quad + \left[\bar{s}\cos\Theta - \bar{d}\sin\Theta\right] M' \left[s\cos\Theta - d\sin\Theta\right] \\ &= \bar{u}Mu + \bar{c}Mc\bar{d}M'd + \bar{s}M's \ , \end{aligned} \tag{1.12}$$

in which all flavour-changing components cancel. Note that this "GIM cancellation" is a direct consequence of the fact that the down-type quarks are rotated by an orthogonal matrix

$$\begin{pmatrix} d' \\ s' \end{pmatrix} = \begin{pmatrix} \cos\Theta & \sin\Theta \\ -\sin\Theta & \cos\Theta \end{pmatrix} \begin{pmatrix} d' \\ s' \end{pmatrix} \tag{1.13}$$

On the basis of this hypothesis, Gaillard and Lee [21] calculated the mass difference in the neutral-Kaon system which depends on the mass of the charm quark, which at that time was hypothetical. Gaillard and Lee estimated the mass of the charm quark to be about 1.5 GeV and published their result in the summer of 1974. The experimental confirmation came in November 1974 with the discovery of narrow resonances in e^+e^- collisions and in proton fixed-target scattering. The resonance found in this famous "November revolution" [22, 23] had a mass of about 3 GeV and was immediately interpreted as a bound state of a charm quark–antiquark pair.

After the discovery of parity violation, it was soon noticed that the combined charge conjugation and parity transformation CP still seemed to be a good discrete symmetry of weak interactions. This belief lasted only until the mid-1960's when Cronin and Fitch discovered CP-violating decays of neutral kaons [24]. Owing to the strangeness quantum number, the neutral kaon cannot be its own antiparticle, and if CP were a good symmetry, these two neutral kaons would combine into two states of definite CP. These neutral kaons decay into either two or three pions, and, again assuming CP symmetry, the CP-even neutral kaon can decay only into two pions, while the CP-odd kaon can decay only into three pions. Since the mass of the kaons just barely allows the decay into three pions, the CP-odd kaon has a much longer lifetime. In their experiment, Cronin and Fitch discovered that the long-lived kaon decayed into two pions in some rare cases, which clearly violates CP.

From the theoretical side, it was clear that the Fermi theory could not be the fundamental theory of weak interactions. When the results were extrapolated to higher energies, it turned out that the cross-sections for $e\nu$ scattering, for example, violated unitarity at energies of the order of 100 GeV. Although this was a gigantic energy at the time Fermi wrote down (1.1), this argument showed that there was a problem at least in principle.

Related to this, it was noticed that the interaction (1.4) allowed only tree-level calculations; any quantum correction turned out to be divergent, and even after the concept of renormalization was developed, Fermi's theory did not have any predictive power once loops were included. From today;s point of view, the corresponding interaction is non-renormalizable and can only be interpreted as an effective interaction valid at very small energies.

It was clear that the high-energy behaviour of Fermi's theory could be improved if, instead of a local interaction, an "intermediate vector boson", which plays the same role as the photon in electromagnetism, was postulated [25]. However, the success of Fermi's theory indicated that such an "intermediate boson" must have a large mass. Naive estimates showed that this mass had to be as large as 100 GeV. Although this improved the high-energy behaviour of the theory, it did not completely solve the unitarity problem of weak interactions. As an example, the scattering of longitudinal "intermediate bosons" still violated unitarity, althought at much larger energies.

The solution of this problem is well known and is only remotely connected to flavour physics. Non-abelian gauge theories, in combination with spontaneous symmetry breakdown, yield a highly predictive framework, certain aspects of which have been tested in detail at LEP. As it has been shown by t'Hooft and Veltman [26], a non-abelian gauge theory with spontaneous symmetry breaking is indeed renormalizable.

Turning again to flavour physics, the Standard Model in its early version contained only two families or, correspondingly, the four quarks mentioned above. It soon became obvious that with only two families and the framework of the Standard Model, CP violation is not possible. It was noticed by Kobayashi and Maskawa in 1974 [27] that CP violation becomes possible in the Standard Model if a third family is postulated. In this case the orthogonal Cabibbo rotation of the down-type quarks is replaced by a unitary rotation, the CKM (Cabibbo, Kobayashi and Maskawa) matrix, yielding an observable phase which allows CP violation.

Subsequently, the particles of the third family were discovered: the τ lepton [28] and the bottom quark [29]. Only the top quark escaped detection for a long time owing to its large mass. While the Standard Model is unable to predict the masses of the particles, a first hint of a possibly very large mass of the top quark came from the observation of $B_0 - \overline{B}_0$ oscillations by ARGUS [30] and UA1 [31], indicating a top-quark mass of well beyond 100 GeV, at a time when the top-quark mass was suspected to be around 25 GeV. Finally, the top quark was discovered in the 1990's at the Tevatron [32].

With the Standard Model, we have today a consistent theory of all particle interactions. The gauge sector of this model has been tested in detail at LEP in the last decade and no significant deviation has been found, despite the fantastic precision of the experiments (see [33] for a review of the LEP results and other results related to electroweak interactions). As far as the flavour sector is concerned, the experiments of the next ten years will show, whether the picture that has developed, in particular the CKM mixing, is correct.

1.2 Importance of Flavour Physics

Understanding flavour mixing in the quark and the leptonic sectors is one of the most important problems of contemporary particle physics. While gauge symmetries provide an elegant way to understand the basic interactions, the sector needed for breaking these gauge symmetries remains a problem, although the Higgs mechanism at least yields consistent quantum field theories. The gauge principle fixes only the interactions of transverse gauge bosons; the nature of the longitudinal polarizations of massive gauge bosons is not yet understood.

The elegance of gauge theories comes from the fact that all interactions are given in terms of a single coupling constant, even when quantum corrections are included. For the Standard Model, this means that all gauge interactions are given in terms of three coupling constants, which can be translated into three parameters: the strong coupling constant α_s, the electromagnetic coupling α_{em} and the weak mixing angle Θ_W. These three parameters correspond to the three factors of the Standard Model gauge group $SU(3)_{\mathrm{colour}} \times SU(2)_{\mathrm{W}} \times U(1)_{\mathrm{Y}}$, each of which introduces a separate gauge coupling. However, in a unified theory based on a simple Lie group, only a single coupling is present in the gauge sector, which means that Θ_{W}, for example, can be computed. The best-known example for this is the $SU(5)$ prediction of the weak mixing angle [34].

Focusing on the Standard Model, this means that only three out of the large number of parameters originate in the gauge sector. Including mixing of the leptons (which implies neutrino masses), the Standard Model has 26 parameters in total, which means that the symmetry-breaking sector induces 23 parameters. Clearly this sector is not as elegant as the gauge sector, since within the Standard Model all these parameters are unrelated.

Reducing the number of parameters in the flavour sector needs physics beyond the Standard Model. In theories with gauge unification, typically the multiplets are larger (containing in general both quarks and leptons) and hence certain relations between masses emerge, such as the famous bottom–τ unification in $SU(5)$ grand unification. However, prediction of the angles and phases of the CKM matrix needs additional input such as symmetries between the families, so-called horizontal symmetries.

Over the next ten years, the sector of the Standard Model related to masses and mixings will be tested experimentally. Hopefully these tests will lead to a hint of what kind of physics beyond the Standard Model is responsible for the flavour structure of the Standard Model. At future experiments such as LHC, precision measurements of B decay will be possible and will lead to a stringent test of the flavour structure of the Standard Model.

1.3 Scope of the Book

There are many excellent textbooks on all of the different aspects of the Standard Model of elementary-particle physics, ranging from the theoretical structure of gauge theories, including their quantization [35], to detailed discussions of the phenomenology of the Standard Model [36, 37, 38, 39, 40, 41], and the present book is not intended to compete with any of these. Rather, it focuses on a special method frequently used in computing the predictions of the Standard Model, which is the method of effective field theory. This approach is well suited to problems involving widely disparate mass scales and hence can even be applied to investigations reaching beyond the Standard Model.

Flavour physics is the sector of the Standard Model which indeed involves widely separated mass scales, and hence effective-field-theory methods are best suited to this field. So it seems worthwhile to collect together the basic ideas of effective field theories and show some of their applications as they appear in the sector of flavour physics.

Quark flavour physics involves scales as high as the weak scale (defined by the weak-boson mass) and as low as Λ_{QCD}, the scale defined by the strong interactions binding the quarks into hadrons. In this sense, it is the ideal field of application of effective field theories. In fact, various effective theories can be constructed: the theory of weak interactions seen at the low scales of weak decays of hadrons is an effective theory ($m_{\mathrm{Hadron}} \ll M_{\mathrm{W}}$), as is the effective theory for heavy quarks ($\Lambda_{\mathrm{QCD}} \ll m_{\mathrm{Quark}}$) and the chiral limit of QCD ($m_{\pi} \ll \Lambda_{\chi\mathrm{SB}}$).

In recent times, lepton flavour physics has also started to become an interesting subject, since neutrino oscillations and thus also neutrino masses seem to have been established by recent experiments. However, the phenomenology of quark flavour physics is currently much richer; this situation will remain for some time, since the B factories will produce data for at least another five years and after that there will be "second–generation" B physics experiments yielding even more precise data. For this reason the emphasis of this book is on quark flavour physics; lepton flavour physics is mentioned only briefly.

The book consists of three parts. After an introduction to flavour in the Standard Model and the CKM mixing matrix, the general ideas of effective field theories are given, followed by discussion of the effective-field-theory approaches used for various purposes in flavour physics. In the subsequent chapters some applications of these methods are considered. Here, we do not aim at completeness; rather, we aim to show how these methods are applied. Finally, the Standard Model itself can be considered an effective field theory, and on this basis one can discuss the physics beyond the Standard Model in general way; we close the book with a few remarks on this point of view.

References

1. H. Becquerel, C. R. Acad. Sci. (Paris) **122**, 501 (1896).
2. E. Rutherford, Phil. Mag. **47**, 109 (1899).
3. J. Chadwick, Verh. Dtsch. Phys. Ges. **16**, 383 (1914).
4. C. D. Ellis and W. A. Wooster, Proc. Roy. Soc. London A **117**, 109 (1927).
5. W. Pauli, *Collected Scientific Papers*, Vol. 2, p. 1313 (Interscience, New York, 1964).
6. F. Reines and C. L. Cowan, Phys. Rev. **92**, 830 (1953).
7. F. Reines, C. L. Cowan, F. B. Harrison, A. D. McGuire and H. W. Kruse, Science **124**, 103 (1956).
8. F. Reines and C. L. Cowan, Phys. Rev. **113**, 273 (1959).
9. E. Fermi, Ricera Scient. **2**, issue 12 (1933); Z. Phys **88**, 161 (1934).
10. G. Gamov and E. Teller, Phys. Rev. **49**, 895 (1936).
11. G. Puppi, Nuovo Cim. **5**, 587 (1948).
12. T. D. Lee and C. N. Yang, Phys. Rev. **104**, 254 (1956).
13. C. S. Wu, E. Ambler, R. Hayward, D. Hoppes and R. Hudson, Phys. Rev. **105**, 1413 (1957).
14. R. L. Garwin, L. M. Lederman and M. Weinrich, Phys. Rev. **105**, 1415 (1957).

15. T. Nakato and K. Nishijima, Prog. Theo. Phys. **10**, 581 (1953).
16. K. Nishijima, Prog. Theor. Phys. **12** 107 (1954), **13**, 285 (1955).
17. M. Gell-Mann, Phys. Rev. **92**, 833 (1953).
18. N. Cabibbo, Phys. Rev. Lett. **10**, 531 (1963).
19. M. Gell-Mann, Phys. Lett. **8**, 214 (1964); Physics **1**, 63 (1964).
20. S. Glashow, J. Iliopoulos and L. Maiani, Phys. Rev. D **2**, 1285 (1970).
21. M. Gaillard and B. Lee, Phys. Rev. D **10**, 897 (1974).
22. J. Aubert et al., Phys. Rev. Lett. **33**, 404 (1974).
23. J. Augustin et al., Phys. Rev. Lett. **33**, 406 (1974).
24. J. Christensen, J. Cronin, V. Fitch and R. Turlay, Phys. Rev. Lett. **13**, 138 (1964); Phys. Rev. **140** B74 (1965).
25. H. Yukawa, Proc. Phys. Math. Soc. Japan **17**, 48 (1935).
26. G. 't Hooft and M. Veltman, Nucl. Phys. B **44**, 189 (1972).
27. M. Kobayashi and T. Maskawa, Progr. Theor. Phys. **49**, 652 (1973).
28. M. Perl et al., Phys. Rev. Lett. **35**, 1489 (1975).
29. S. Herb et al., Phys. Rev. Lett. **39**, 252 (1977).
30. H. Albrecht et al., Phys. Lett. **192B**, 245 (1987).
31. C. Albajar et al., Phys. Lett. **186B**, 247 (1987).
32. F. Abe et al. [CDF Collaboration], Phys. Rev. Lett. **73**, 225 (1994) [arXiv:hep-ex/9405005].
33. P. Wells, "Experimental Tests of the Standard Model", plenary talk at EPS2003, Aachen, 17–23 July 2003.
34. H. Georgi and S. L. Glashow, Phys. Rev. Lett. **32**, 438 (1974).
35. L. D. Faddeev and A. A. Slavnov, *Gauge Fields. Introduction To Quantum Theory*, Frontiers in Physics, No. 83, (Addison-Wesley, Redwood City,1990).
36. C. Quigg, *Gauge Theories of the Strong, Weak and Electromagnetic Interactions*, Frontiers in Physics, No. 56, (Addison-Wesley, Redwood City, 1983).
37. K. Huang, *Quarks Leptons and Gauge Fields* (World Scientific, Singapore, 1992).
38. P. Renton, *Electroweak Interactions* (Cambridge University Press, Cambridge, 1990)
39. J. Donoghue, E. Golowich, B. Holstein, *Dynamics of the Standard Model*, Cambridge Monographs on Particle Physics (Cambridge University Press, Cambridge, 1992)
40. O. Nachtmann, *Elementary Particle Physics*, Springer Texts and Monographs in Physics (Springer, Berlin, Heidelberg, 1989).
41. M. Böhm, A. Denner and H. Joos, *Gauge Theories of the Strong and Electroweak Interaction* (Teubner, Stuttgart, 2001)

2 Flavour in the Standard Model

2.1 Basics of the Standard Model

All known phenomenology of elementary particles can be described in terms of the so-called Standard Model [1, 2, 3, 4, 5, 6, 7], which has turned out to be an extraordinarily successful theory. It describes all known phenomenology from very low scales up to the highest experimentally accessible scales. Certain aspects of the Standard Model, namely the couplings of the Z_0 gauge boson to the fermions, have been tested at a level of precision well below 1%, and no significant deviation has been found.

The Standard Model is constructed as a spontaneously broken $SU(3)_{\text{colour}} \times SU(2)_{\text{W}} \times U(1)_{\text{Y}}$ gauge theory [8, 9, 10, 11, 12, 13], where the $SU(3)_{\text{colour}}$ corresponds to the strong interaction and the $SU(2)_{\text{W}} \times U(1)_{\text{Y}}$ induces the electroweak interaction. The gauge group has 12 generators, corresponding to eight gluons g for the strong interaction, three weak bosons $W^{\pm}$ and Z^0, and the photon mediating the electromagnetic interaction.

The matter fields, i.e. the quarks and leptons, have to be grouped into multiplets of the gauge group, i.e. they have to be assigned electroweak and strong quantum numbers. Parity violation in weak interactions is implemented by assigning different weak quantum numbers to left- and right-handed components of the matter fields. In other words, the left- and right-handed components of the quarks and leptons are associated with different multiplets of the electroweak $SU(2)_{\text{W}} \times U(1)_{\text{Y}}$ group. The left-handed leptons are grouped into doublets of $SU(2)$ in the following way:

$$L_e = \begin{pmatrix} \nu_{e,\text{L}} \\ e_{\text{L}} \end{pmatrix} , \quad L_m = \begin{pmatrix} \nu_{\mu,\text{L}} \\ \mu_{\text{L}} \end{pmatrix} , \quad L_t = \begin{pmatrix} \nu_{\tau,\text{L}} \\ \tau_{\text{L}} \end{pmatrix} , \tag{2.1}$$

where the subscript L means the left-handed projection of the spinor fields

$$\psi_{\text{L}} = \frac{1}{2}(1 - \gamma_5)\psi .$$

Similarly, for the quarks the assignment is

$$Q_d = \begin{pmatrix} u_{\text{L}} \\ d_{\text{L}} \end{pmatrix} , \quad Q_s = \begin{pmatrix} c_{\text{L}} \\ s_{\text{L}} \end{pmatrix} , \quad Q_b = \begin{pmatrix} t_{\text{L}} \\ b_{\text{L}} \end{pmatrix} . \tag{2.2}$$

A transformation Λ of $SU(2)_L$ is a unitary 2×2 matrix and these doublets transform as

$$L_i' = \Lambda L_i \ , \quad Q_i' = \Lambda Q_i \ , \qquad \text{for } \Lambda \in SU(2)_{\mathrm{L}} \ . \tag{2.3}$$

In order to introduce mass terms, one has also to have right-handed components of the spinor fields, since a mass terms corresponds to a coupling term between right- and left-handed components. As far as the weak $SU(2)_{\mathrm{W}}$ group is concerned, the right-handed components transform as singlets under this group; in other words, they do not couple to the gauge bosons corresponding to $SU(2)_{\mathrm{W}}$.

However, as we shall see below, the Higgs sector of the Standard Model has in fact a larger symmetry, which is an $SU(2)_{\mathrm{L}} \times SU(2)_{\mathrm{R}}$ symmetry. This so-called custodial symmetry [14, 15, 16] is broken by the quark mass terms and also by the gauge couplings, but it plays a role in unified models.

In anticipation of the discussion of custodial symmetry, it is useful to group the right-handed quarks and leptons also into doublets, of a group $SU(2)_{\mathrm{R}}$. Thus we write

$$\ell_1 = \begin{pmatrix} \nu_{e,\mathrm{R}} \\ e_{\mathrm{R}} \end{pmatrix} \ , \quad \ell_2 = \begin{pmatrix} \nu_{\mu,\mathrm{R}} \\ \mu_{\mathrm{R}} \end{pmatrix} \ , \quad \ell_3 = \begin{pmatrix} \nu_{\tau,\mathrm{R}} \\ \tau_{\mathrm{R}} \end{pmatrix} \tag{2.4}$$

for the right-handed leptons and

$$q_1 = \begin{pmatrix} u_{\mathrm{R}} \\ d_{\mathrm{R}} \end{pmatrix} \ , \quad q_2 = \begin{pmatrix} c_{\mathrm{R}} \\ s_{\mathrm{R}} \end{pmatrix} \ , \quad q_3 = \begin{pmatrix} t_{\mathrm{R}} \\ b_{\mathrm{R}} \end{pmatrix} \tag{2.5}$$

for the right-handed quarks. A transformation of $SU(2)_{\mathrm{R}}$ is a unitary 2×2 matrix R and the transformation for the doublets is

$$\ell_i' = R L_i \ , \quad q_i' = R Q_i \ , \qquad \text{for } R \in SU(2)_{\mathrm{R}} \ . \tag{2.6}$$

Only the left-handed group $SU(2)_{\mathrm{L}} \equiv SU(2)_{\mathrm{W}}$ is gauged, and yields the usual couplings of the gauge bosons to the quarks and leptons. The hypercharge group $U(1)_{\mathrm{Y}}$ has to be identified with a combination of the phase transformation of the fields and a transformation in the $T_{3,\mathrm{R}}$ direction of $SU(2)_{\mathrm{R}}$. Consequently, the right-handed $SU(2)_{\mathrm{R}}$ is broken by the hypercharge gauge coupling and, as we shall see later, by the mass terms. The hypercharge assignment is determined by the requirement that the particles should have the correct charge. For the leptons we obtain

$$Y = -1 + 2T_{3,\mathrm{R}} \ , \tag{2.7}$$

while for the quarks we obtain

$$Y = \frac{1}{3} + 2T_{3,\mathrm{R}} \ . \tag{2.8}$$

The charge $\mathcal{Q}$ of these particles is obtained from

$$\mathcal{Q} = \frac{1}{2}\left(2T_{3,L} + Y\right) . \tag{2.9}$$

The hypercharge assignments for quarks and leptons can be written in the form

$$Y = B - L + 2T_{3,\mathrm{R}} , \tag{2.10}$$

where B is the baryon number and L is the lepton number of the state. The relation (2.10) plays a role in unified theories, where typically quarks and leptons appear in the same multiplet.

With these assignments, all couplings to the gauge bosons of the electroweak interactions are fixed. Furthermore, as far as the strong $SU(3)_\mathrm{C}$ group is concerned, all leptons are singlets and all quarks (left- and right-handed) are triplets, fixing also the coupling to the gluons via the gauge principle.

Since we are dealing with a chiral gauge theory (i.e. left- and right-handed components have different quantum numbers), the symmetry forbids mass terms as long as it is unbroken, except for the right-handed neutrino, which carries neither $SU(2)_\mathrm{L}$ quantum numbers nor a hypercharge. In this case a Majorana mass term is allowed, which we shall discuss in Sect. 2.3. All other particles have to obtain their mass from symmetry breaking which we shall discuss in the next section.

2.2 The Higgs Sector and Yukawa Couplings

It is interesting to note that the complete flavour structure of the Standard Model is fixed by the Yukawa couplings of the quarks and leptons to the Higgs sector. Furthermore, the fact that $SU(2)_\mathrm{L} \equiv SU(2)_\mathrm{W}$ and $U(1)_\mathrm{Y}$ are gauged seems to be irrelevant for the flavour structure.

To discuss these issues, we start from the particle doublets (2.1), (2.2), (2.4) and (2.5) and consider first the quarks. We can write a kinetic energy for the quarks as

$$\mathcal{L}_\mathrm{kin} = \sum_i \left[\bar{Q}_i \partial\!\!\!/ Q_i + \bar{q}_i \partial\!\!\!/ q_i\right] , \tag{2.11}$$

which is symmetric under $U(2)_\mathrm{L} \times U(2)_\mathrm{R}$. A mass term would break this symmetry explicitly down to the diagonal symmetry $SU(2)_\mathrm{L+R}$ (i.e. the transformation of $SU(2)_\mathrm{L}$ has to be chosen to be equal to that of $SU(2)_\mathrm{R}$), but let us first maintain the larger symmetry.

In addition to the fermion fields we introduce a set of scalar fields, gathered into a 2×2 matrix

$$H = \begin{pmatrix} \phi_0 + i\chi_0 & \sqrt{2}\phi_+ \\ -\sqrt{2}\phi_- & \phi_0 - i\chi_0 \end{pmatrix} , \tag{2.12}$$

where ϕ_0 and χ_0 are real fields, and $\phi_+^* = \phi_-$ is a complex field. The transformation properties of this matrix are

$$H \to \Lambda H R^\dagger \quad \text{for } \Lambda \in SU(2)_\mathrm{L}\,, \quad R \in SU(2)_\mathrm{R}\,. \tag{2.13}$$

With the help of this field, we can write a Lagrangian which is invariant under $SU(2)_L \times SU(2)_R$. The part for the scalar fields reads

$$\mathcal{L}_\mathrm{Higgs} = \frac{1}{4}\mathrm{Tr}\left[(\partial_\mu H)^\dagger(\partial^\mu H)\right] - V(\mathrm{Tr}\left[H^\dagger H\right])\,, \tag{2.14}$$

where the Higgs potential V will be discussed below. The only possible renormalizable and $SU(2)_\mathrm{L} \times SU(2)_\mathrm{R}$-invariant interaction between the scalar fields and the quarks is

$$\mathcal{L}_I = -\sum_{ij} y_{ij}\bar{Q}_i H q_j + \text{ h.c. }\,, \tag{2.15}$$

where y is the 3×3 matrix of coupling constants.

The total Lagrangian is the sum of the terms (2.11), (2.14) and (2.15). It has an $SU(2)_L \times SU(2)_R$ symmetry and is basically the Lagrangian of the linear σ model [17]. The matrix y of Yukawa couplings can be diagonalized by a bi-unitary transformation

$$y = U^\dagger y_\mathrm{diag} W\,, \tag{2.16}$$

with two unitary 3×3 matrices U and W. Redefining left- and right-handed quarks appropriately, we find that the total $SU(2)_\mathrm{L} \times SU(2)_\mathrm{R}$-invariant Lagrangian is diagonal as far as the family structure is concerned, i.e.

$$\begin{aligned}\mathcal{L} = \sum_i \left[\bar{Q}_i \partial\!\!\!/ Q_i + \bar{q}_i \partial\!\!\!/ q_i\right] + \frac{1}{4}\mathrm{Tr}\left[(\partial_\mu H)^\dagger(\partial^\mu H)\right] - V(\mathrm{Tr}\left[H^\dagger H\right]) \\ -\sum_i y_i \bar{Q}_i H q_i + \text{ h.c. }\,,\end{aligned} \tag{2.17}$$

and hence no mixing between different quark families can occur.

The Higgs potential is chosen in such a way that the field H acquires a vacuum expectation value, which can be chosen as

$$\langle\phi_0\rangle = v \text{ or } \langle H\rangle = v\,1_{2\times 2} \tag{2.18}$$

such that

$$H = \begin{pmatrix} v + h_0 + i\chi_0 & \sqrt{2}\phi_+ \\ -\sqrt{2}\phi_- & v + h_0 - i\chi_0 \end{pmatrix}\,. \tag{2.19}$$

This vacuum expectation value breaks $SU(2)_\mathrm{L} \times SU(2)_\mathrm{R}$ down to the diagonal $SU(2)_\mathrm{L+R}$ symmetry, which will be discussed below. The resulting spectrum contains a massive Higgs boson h_0 and three massless Goldstone bosons [18] (χ_0, ϕ_+ and ϕ_-), which is typical for spontaneous breakdown. Under $SU(2)_{L+R}$ h_0 is a a singlet, while χ_0, ϕ_+ and ϕ_- form a triplet.

This induces mass terms for the quarks, which originate from the Yukawa couplings. We obtain

$$\begin{aligned}\mathcal{L}_{\text{mass}} &= -\sum_i y_i v \bar{Q}_i q_i + \text{ h.c.} \\ &= -m_u(\bar{u}u + \bar{d}d) - m_c(\bar{c}c + \bar{s}s) - m_c(\bar{t}t + \bar{b}b) \ .\end{aligned} \tag{2.20}$$

Clearly, the quark mass spectrum of this Lagrangian is phenomenologically not acceptable. This is due to the fact that the symmetry of this Lagrangian is larger than what is actually needed for the Standard Model. The hypercharge of the Standard Model involves $T_{3,\mathrm{R}}$, one of the generators of $SU(2)_{\mathrm{R}}$. Thus we can explicitly break $SU(2)_{\mathrm{R}}$ with terms proportional to $T_{3,\mathrm{R}}$ without violating the symmetries of the Standard Model.

In the Higgs Lagrangian (2.14), we can introduce $T_{3,\mathrm{R}}$ contributions by considering

$$\mathrm{Tr}\left[(\partial_\mu H)^\dagger(\partial^\mu H)T_{3,R}\right] \quad \text{and} \quad V'(\mathrm{Tr}\left[H^\dagger H T_{3,R}\right]) \ , \tag{2.21}$$

but these terms vanish or yield an irrelevant constant owing to the relation $\mathrm{Tr}\left[H^\dagger H T_{3,R}\right] = 0$. This means that the Standard Model Higgs sector automatically has the larger $SU(2)_{\mathrm{L}} \times SU(2)_{\mathrm{R}}$ symmetry once one implements the $SU(2)_{\mathrm{L}} \times U(1)_{\mathrm{Y}}$ symmetry of the Standard Model.

This custodial symmetry [14, 15, 16] is specific to the breaking of the $SU(2)_{\mathrm{L}} \times U(1)_{\mathrm{Y}}$ symmetry by a doublet of scalar fields. The vacuum expectation value of the Higgs field is proportional to the 2×2 unit matrix and thus is invariant under the diagonal $SU(2)_{L+R}$ group. Thus, after symmetry breaking, the Higgs sector still has an unbroken custodial $SU(2)$ symmetry, under which the three Goldstone bosons transform as a triplet and the physical Higgs boson transforms as a singlet. After $SU(2)_{\mathrm{L}}$ is gauged, the massless Goldstone bosons become the longitudinal modes of the gauge bosons, and thus the three gauge bosons are also a triplet under custodial $SU(2)$.

This symmetry has has some interesting consequences. Since the gauge bosons form a triplet under custodial $SU(2)$, the strengths of charged and neutral currents have to be equal. The ratio of these coupling strengths is called the ρ parameter, which is fixed at unity in the symmetry limit. Furthermore, exact custodial $SU(2)$ would enforce equal up and down quark masses within one family and it would forbid quark flavour mixing.

However, the quark Yukawa couplings break custodial $SU(2)$; we can write an additional Yukawa coupling term that explicitly breaks custodial $SU(2)$,

$$\mathcal{L}'_I = -\sum_{ij} y'_{ij} \bar{Q}_i H T_{3,R} q_j + \text{ h.c.} \ , \tag{2.22}$$

which will lead to both family mixing and a mass splitting of the up and down quark masses within one family, since y and y' cannot be diagonalized simultaneously.

For the following discussion, it is useful to introduce three-component objects in the form

$$\mathcal{U}_{\mathrm{L/R}} = \begin{bmatrix} u_{\mathrm{L/R}} \\ c_{\mathrm{L/R}} \\ t_{\mathrm{L/R}} \end{bmatrix} , \quad \mathcal{D}_{\mathrm{L/R}} = \begin{bmatrix} d_{\mathrm{L/R}} \\ s_{\mathrm{L/R}} \\ b_{\mathrm{L/R}} \end{bmatrix} . \tag{2.23}$$

The total Lagrangian, consisting of (2.17) and (2.22), can be rewritten in terms of (2.23) and reads

$$\begin{aligned} \mathcal{L} = {} & \bar{\mathcal{U}}_L \partial\!\!\!/ \mathcal{U}_L + \bar{\mathcal{U}}_R \partial\!\!\!/ \mathcal{U}_R + \bar{\mathcal{D}}_L \partial\!\!\!/ \mathcal{D}_L + \bar{\mathcal{D}}_R \partial\!\!\!/ \mathcal{D}_R \\ & + (\partial\phi_+)(\partial\phi_-) + \frac{1}{2}(\partial\phi_0)(\partial\phi_0) + \frac{1}{2}(\partial\chi_0)(\partial\chi_0) - V(2\phi_+\phi_- + \phi_0^2 + \chi_0^2) \\ & + \left[\frac{1}{v}\bar{\mathcal{U}}_L \mathcal{M}_u \phi_0 \mathcal{U}_R + \text{ h.c.}\right] + \left[\frac{1}{v}\bar{\mathcal{D}}_L \mathcal{M}_d \phi_0 \mathcal{D}_R + \text{ h.c.}\right] \\ & + \left[\frac{i}{v}\bar{\mathcal{U}}_L \mathcal{M}_u \chi_0 \mathcal{U}_R + \text{ h.c.}\right] - \left[\frac{i}{v}\bar{\mathcal{D}}_L \mathcal{M}_d \chi_0 \mathcal{D}_R + \text{ h.c.}\right] \\ & + \left[\frac{1}{v}\bar{\mathcal{D}}_L \mathcal{M}_u \phi_+ \mathcal{U}_R + \text{ h.c.}\right] + \left[\frac{1}{v}\bar{\mathcal{U}}_L \mathcal{M}_d \phi_- \mathcal{D}_R + \text{ h.c.}\right] , \end{aligned} \tag{2.24}$$

where we have defined the 3×3 mass matrices for the up and down quarks as

$$\mathcal{M}_u = v(y + y') , \quad \mathcal{M}_d = v(y - y') . \tag{2.25}$$

The somewhat lengthy expression (2.24) is the full Higgs and Yukawa sector of the Standard Model, containing its full flavour structure for the quarks. After spontaneous symmetry breaking, the field ϕ_0 acquires a vacuum-expectation value in accordance with (2.18); this corresponds to the replacement $\phi_0 \to v + \phi_0$. The fields $\phi_\pm$ are massless and become the longitudinal components of the charged gauge bosons, while the massless field χ_0 becomes the longitudinal mode of the neutral boson. We shall no present the details of the Higgs mechanism here; rather, we refer the reader to textbooks [8, 9, 10, 11, 12, 13, 19].

Mixing between different families occurs through the fact that the two mass matrices $\mathcal{M}_u$ and $\mathcal{M}_d$ not commute any more, i.e.

$$[\mathcal{M}_u , \mathcal{M}_d] \neq 0 , \tag{2.26}$$

which is a direct consequence of the explicit breaking of custodial $SU(2)$ symmetry through the Yukawa couplings of the quarks. The mixing between different quark families is encoded in the CKM matrix, which will be discussed in the next chapter.

In summary, the structure of the Higgs sector of the Standard Model is completely equivalent to the σ-model [17] which was invented in a totally different context long before the construction of the Standard Model. The way we have presented our derivation up to this point corresponds to the *linear* σ model, which means that the $SU(2) \times SU(2)$ symmetry is realized linearly. We shall later also use the *non-linear* σ model, which is formally obtained in the limit in which the mass of the physical Higgs boson h_0 tends to infinity. In this limit, this particle decouples and only the three Goldstone bosons remain. Formally, this is obtained by the replacement

$$H \to v\Sigma \,, \tag{2.27}$$

where Σ has the same transformation properties as H but contains only the three Goldstone boson fields. The $SU(2) \times SU(2)$ symmetry is realized on these three fields in a non-linear way [20, 21]. The Higgs contribution to the Lagrangian (2.17) simplifies, since

$$\Sigma\Sigma^\dagger = \Sigma^\dagger\Sigma = 1 \tag{2.28}$$

such that

$$\mathcal{L}_{\text{Higgs}} = \frac{v^2}{4}\text{Tr}\left[(\partial_\mu\Sigma)^\dagger(\partial^\mu\Sigma)\right] \,, \tag{2.29}$$

since the potential becomes an irrelevant constant. Note that the discussion of the masses and mixings of the quarks remains the same independent of which representation (linear or non-linear σ model) is chosen for the Higgs sector.

We have not yet introduced the gauge fields for the strong, weak and electromagnetic interactions. However, in order to understand the flavour structure of the Standard Model, *these fields are not needed.* In other words, the flavour physics in the Standard Model originates completely in the scalar sector responsible for the breaking of the electroweak $SU(2)\times U(1)$ symmetry, since up to now we have used this symmetry only as a spontaneously broken *global* symmetry. On the other hand, the spontaneous breakdown of a global symmetry implies the appearance of massless Goldstone bosons, which is phenomenologically not acceptable. To avoid the appearance of these states, one can use the Higgs mechanism [4, 5, 6, 7] to turn them into longitudinal modes of massive gauge bosons. We shall return to this point when we discuss Fermi's theory of weak interactions as an effective theory.

2.3 Neutrino Masses and Lepton Mixing

The leptonic sector can, in large part, be treated along the same lines. It has been assumed until recently that neutrinos are massless, and in this case no right-handed components are needed for those particles. This has the consequence that all the rotation matrices needed to diagonalize the Yukawa couplings can be rotated away such that no family mixing occurs in the leptonic sector. In other words, in the case of massless neutrinos, separate lepton numbers for the electron, the muon and the τ lepton exist, which are conserved.

However, there has been recent evidence that neutrinos have masses and consequently also mixing [22, 23, 24]. It is still quite possible that the leptonic sector is just a copy of what happens in the quark sector, but now neutrinos have to have right-handed components, and a mixing matrix similar to the CKM matrix appears. Grouping the leptons into doublets as in (2.1) and (2.4), we can go through the same steps as for the quarks and obtain a very similar structure.

There is, however, the possibility that the leptonic sector is different from the quark sector in the following respect [25]. Looking at the relation for the hypercharge of the leptons (2.7), we find that the right-handed neutrino carries neither hypercharge nor weak $SU(2)_\mathrm{L}$ charge, i.e. the right-handed neutrino does not carry any $U(1)$ charge. Thus we may assume that the right-handed neutrino is equal to its antiparticle, i.e. we may assume that the right-handed neutrino is a Majorana fermion. In this case one can write a Majorana mass term for the right-handed neutrinos; this mass term does not come from the Yukawa couplings to the Higgs field.

In order to write such a mass term, we observe that the charge conjugate-field of a right-handed fermion is left-handed. Using the usual definition of charge conjugation (see also Sect. 6.1)

$$\psi^\mathrm{c} = C\bar{\psi}^\mathrm{T} , \qquad \text{where } C = -i\gamma_0\gamma_2 , \tag{2.30}$$

we can write a mass term for the right handed neutrinos of the form

$$\mathcal{L}_\mathrm{MM} = -\frac{1}{2}\bar{\nu}_{\mathrm{R},i} M_{ij} \nu^\mathrm{c}_{\mathrm{R},j} + \text{ h.c.} , \tag{2.31}$$

where i and j are now indices for the three families. The matrix M_{ij} has to be symmetric and is called the Majorana mass matrix. Note that this mass term violates lepton number, since it carries two units of lepton number.

From the usual couplings with the Higgs field we obtain another mass term for the neutrinos which is the usual Dirac mass term. This Dirac mass term can be written as

$$\mathcal{L}_{DM} = -\bar{\nu}_{L,i} m_{ij} \nu_{R,j} + \text{h.c.} \tag{2.32}$$

and is obtained from the coupling of the lepton doublets to the Higgs field. The complete mass term can thus be written as

$$\mathcal{L}_\mathrm{M} = -\frac{1}{2}\left(\bar{\nu}_{\mathrm{L},i}\ \overline{(\nu^\mathrm{c}_\mathrm{R})}_{\mathrm{L},i}\right)\begin{pmatrix} 0 & m_{ij} \\ m^\mathrm{T}_{ij} & M_{ij}\end{pmatrix}\begin{pmatrix}(\nu^\mathrm{c}_\mathrm{L})_{\mathrm{R},j} \\ \nu_{\mathrm{R},j}\end{pmatrix} , \tag{2.33}$$

where we have made use of the relation

$$\overline{(\nu^\mathrm{c}_\mathrm{R})}_\mathrm{L}(\nu^\mathrm{c}_\mathrm{L})_\mathrm{R} = \bar{\nu}_\mathrm{L}\bar{\nu}_\mathrm{R} \tag{2.34}$$

and introduced a 6×6 matrix, which has the particular block structure indicated in (2.33).

The fact that right-handed neutrinos do not interact except through the Lagrangian $\mathcal{L}_{\mathrm{M}}$ offers an interesting possibility of generating small neutrino masses using the so-called see-saw mechanism. Since the Majorana mass term (2.31) is not due to the Higgs mechanism, there is no connection to the electroweak vacuum expectation value. Thus the Majorana masses of the right-handed neutrinos can in principle be large, maybe even as large as the scale of grand unification. In this case we can integrate out the right-handed neutrinos and study the higher-dimensional operators induced by this operation. In practical terms, this means that we can replace all right-handed neutrino fields in the interaction terms with all other fields by

$$\nu_{\mathrm{R}_i} = M_{ij}^{-1} m_{jk} \nu_{\mathrm{L},k} \,, \tag{2.35}$$

which is the equation of motion for small momenta, i.e. he equation obtained by neglecting the kinetic energy of the right-handed neutrino.

The main effect of this is that a dimension-five operator appears which introduces a Majorana mass terms for the left-handed neutrinos of the form

$$\mathcal{L}'_{\mathrm{MM}} = -\frac{1}{2}\left(\nu_{\mathrm{L},i}^{\mathrm{T}}\left[m^{\mathrm{T}} M^{-1} m\right]_{ij} C \nu_{\mathrm{L},j} + \bar{\nu}_{\mathrm{L},i}\left[m^{\mathrm{T}} M^{-1} m\right]_{ij} C \bar{\nu}_{\mathrm{L},i}^{\mathrm{T}}\right) \,, \tag{2.36}$$

where now the Majorana masses of the left-handed neutrinos are small, since they are suppressed by the large Majorana masses of the right-handed neutrinos. This see-saw mechanism was discussed first in [25, 26] and offers a natural way to obtain the small observed neutrino masses.

The mass matrices in (2.36) are still not diagonal; in order to diagonalize the mass matrices one has to perform a rotation, which in this case is now an orthogonal transformation, since the mass matrix is now symmetric:

$$m^{\mathrm{T}} M^{-1} m = \mathcal{O}^{\mathrm{T}} \mu_{diag} \mathcal{O} \,, \tag{2.37}$$

where μ_{diag} is the diagonal matrix with the mass eigenvalues of the left-handed neutrinos.

The masses of the charged leptons are generated in the same way as for the down-type quarks; the resulting mass matrix $\mathcal{M}_L$ is diagonalized by a bi-unitary transformation

$$\mathcal{M}_{\mathrm{L}} = U^{\dagger} \mathcal{M}_{\mathrm{L,diag}} W \,. \tag{2.38}$$

As in the quark case, the effect of these rotations can be observed in the charged current only, where a mixing matrix similar to the CKM matrix appears. The charged-current interaction in the mass eigenbasis reads

$$\mathcal{L}_{\mathrm{CC}} = \frac{g}{\sqrt{2}} \bar{\nu}_{\mathrm{L}} \gamma^{\mu} V_{\mathrm{MNS}} l_{\mathrm{L}} W_{\mu}^{+} \tag{2.39}$$

where l_L are the three left-handed charged leptons, and

$$V_{\mathrm{MNS}} = U \mathcal{O}^{\mathrm{T}} \tag{2.40}$$

is the unitary Maki–Nakagawa–Sakata matrix [27].

The counting of parameters is, however, slightly different from the case of quarks. Since the left-handed neutrinos are now Majorana fermions, there is no more freedom to rephase these fields. For n families, the unitary MNS matrix again has n^2 real parameters, but we have now the freedom to rephase the n charged leptons, i.e. we may choose their n phases relative to the left-handed neutrinos. When this has been done, we have $n(n-1)$ free real parameters, of which $n(n-1)/2$ can be interpreted as the Euler angles of an orthogonal rotation. The remaining $n(n-1)/2$ parameters are (irreducible) phases, which lead to CP violation in the leptonic sector. One of these phases is similar to that which appears in the CKM matrix and can be observed by comparing the oscillation rates $P(\nu_i \to \nu_j)$ for neutrinos with the corresponding rates for antineutrinos $P(\bar{\nu}_i \to \bar{\nu}_j)$. The other two phases are related to the Majorana nature of the neutrino and are very difficult to extract from a measurement.

As stated above, the presence of the Majorana mass terms violates lepton number. After the heavy right-handed neutrino is integrated out, a Majorana the mass term (2.36) appears for the light neutrinos, implying lepton number violation. However, this contribution is suppressed by the large scale of the Majorana mass of the right-handed neutrinos and hence this effect is very small. We shall not go into any more detail concerning this subject and refer the reader instead to dedicated textbooks on neutrino physics such as [28].

References

1. S. L. Glashow, Nucl. Phys. **22**, 579 (1961).
2. S. Weinberg, Phys. Rev. Lett. **19**, 1264 (1967).
3. A. Salam, proceedings of the Nobel Symposium, Stockholm, 1968, p. 367.
4. P. W. Higgs, Phys. Rev. Lett. **13**, 508 (1964).
5. P. W. Higgs, Phys. Rev. **145**, 1156 (1966).
6. G. S. Guralnik, C. R. Hagen and T. W. B. Kibble, Phys. Rev. Lett. **13**, 585(1964).
7. T. W. B. Kibble, Phys. Rev. **155**, 1554 (1967).
8. C. Quigg, *Gauge Theories of the Strong, Weak and Electromagnetic Interactions*, Frontiers in Physics, No. 56, (Addison-Wesley, Redwood City, 1983).
9. K. Huang, *Quarks Leptons and Gauge Fields* (World Scientific, Singapore, 1992).
10. P. Renton, *Electroweak Interactions* (Cambridge University Press, Cambridge, 1990)
11. J. Donoghue, E. Golowich, B. Holstein, *Dynamics of the Standard Model*, Cambridge Monographs on Particle Physics (Cambridge University Press, Cambridge, 1992)
12. O. Nachtmann, *Elementary Particle Physics*, Springer Texts and Monographs in Physics (Springer, Berlin, Heidelberg, 1989).
13. M. Böhm, A. Denner and H. Joos, *Gauge Theories of the Strong and Electroweak Interaction* (Teubner, Stuttgart, 2001)

14. M. J. G. Veltman, Nucl. Phys. B **123**, 89 (1977).
15. P. Sikivie, L. Susskind, M. B. Voloshin and V. I. Zakharov, Nucl. Phys. B **173**, 189 (1980).
16. M. S. Chanowitz, M. A. Furman and I. Hinchliffe, Phys. Lett. B **78**, 285 (1978).
17. M. Gell-Mann and M. Levy, Nuovo Cim. **16**, 705 (1960).
18. J. Goldstone, Nuovo Cim. **19**, 154 (1961).
19. L. D. Faddeev and A. A. Slavnov, *Gauge Fields. Introduction To Quantum Theory*, Frontiers in Physics, No. 83, (Addison-Wesley, Redwood City,1990).
20. S. R. Coleman, J. Wess and B. Zumino, Phys. Rev. **177**, 2239 (1969).
21. C. G. Callan, S. R. Coleman, J. Wess and B. Zumino, Phys. Rev. **177** (1969) 2247.
22. Y. Fukuda et al. [Super-Kamiokande Collaboration], Phys. Rev. Lett. **81**, 1562 (1998) [arXiv:hep-ex/9807003].
23. Q. R. Ahmad et al. [SNO Collaboration], Phys. Rev. Lett. **87**, 071301 (2001) [arXiv:nucl-ex/0106015].
24. Q. R. Ahmad et al. [SNO Collaboration], Phys. Rev. Lett. **89**, 011301 (2002) [arXiv:nucl-ex/0204008].
25. T. Yanagida, in *Proceedings of the Workshop on Unified Theory and Baryon Number of the Universe*, eds. O. Swada and A. Sugamoto, p. 95 (KEK, Tsukuba, 1979).
26. M. Gell-Mann, P. Rammond and R. Slansky, in *Supergravity*, eds. P. van Niewenhuizen and D. Freedman (North-Holland, Amsterdam, 1979).
27. Z. Maki, M. Nakagawa and S. Sakata, Prog. Theor. Phys. **28**, 870 (1962).
28. N. Schmitz, *Neutrino Physik* [in German], (Teubner, 1997).

3 The CKM Matrix and CP Violation

3.1 The CKM Matrix in the Standard Model

In the Standard Model and in all theories with gauge unification, the CKM matrix originates from the fact that the mass matrices of the up and down quarks do not commute (see (2.26)). This means that there is no basis in family space where both matrices are diagonal. The CKM matrix emerges, from this point of view, as the rotation between the two eigenbases of the up and down mass matrices.

We can first redefine the quark fields in such a way that both mass matrices are Hermitian. Furthermore, since only the relative orientation of the two bases is observable, we can perform an unobservable rotation which diagonalizes the up mass matrix; thus we can, without restriction of generality, write

$$\mathcal{M}_u = \begin{pmatrix} m_u & 0 & 0 \\ 0 & m_c & 0 \\ 0 & 0 & m_t \end{pmatrix} , \tag{3.1}$$

where m_u, m_c and m_t are real, positive entries.

In this basis the down mass matrix has to be diagonalized by a non-trivial rotation. Since the matrix is Hermitian, this can be done by a unitary transformation, such that

$$\mathcal{M}_d = V_{\mathrm{CKM}} \mathcal{M}_d^{\mathrm{diag}} V_{CKM}^{\dagger} , \tag{3.2}$$

where

$$\mathcal{M}_d^{\mathrm{diag}} = \begin{pmatrix} m_d & 0 & 0 \\ 0 & m_s & 0 \\ 0 & 0 & m_b \end{pmatrix} , \tag{3.3}$$

again with real, positive entries. The unitary rotation that transforms the eigenbasis of the up-quark mass matrix $\mathcal{M}_u$ into that of the down-quark mass matrix is called Cabibbo–Kobayashi–Maskawa matrix.[1]

Usually the fields in the Lagrangian are interpreted in terms of mass eigenstates which means that one has to redefine the fields in such a way

[1] We could equally well have started from a basis in which the down-quark mass matrix is diagonal. This would lead to the same result, i.e. $V_{CKM}^{\dagger}$ would diagonalize the up-quark mass matrix.

that the mass matrices are diagonal. This means that we have to redefine all down-quark fields as

$$\mathcal{D}_{\mathrm{L/R}} \to V_{\mathrm{CKM}}\mathcal{D}_{\mathrm{L/R}} \,. \tag{3.4}$$

This unitary rotation makes the mass matrices $\mathcal{M}_u$ and $\mathcal{M}_d$ diagonal.

However, this rotation affects the other terms in the Lagrangian. The kinetic energy is invariant under a unitary redefinition of the fields. Likewise, since the neutral currents, i.e. the interactions with the fields ϕ_0 and χ_0, are also invariant under a rotation of the down quarks, there will be no flavour-changing neutral currents in the Standard Model, at least at tree level. This is the modern implementation of the GIM mechanism [1] discussed in Chap. 1

As we shall see later, loop processes will induce flavour-changing neutral currents. However, either the corresponding loop diagrams are convergent or the divergences cancel between different contributions. Consequently, no renormalizing counterterms will be induced as a tree-level contribution and thus the structure of the Lagrangian is preserved even at loop level. Phenomenologically, this means that in the quantum field theory the processes involving flavour-changing neutral currents remain suppressed by small couplings and loop factors.

Only in the charged currents connecting up with down quarks will a visible effectl occur, namely

$$\mathcal{L}_{\mathrm{CC}} = \frac{1}{v}\left[\bar{\mathcal{D}}_{\mathrm{L}} V_{\mathrm{CKM}}^{\dagger}\mathcal{M}_u\phi_+\mathcal{U}_{\mathrm{R}} + \bar{\mathcal{U}}_{\mathrm{L}}\mathcal{M}_d V_{\mathrm{CKM}}\phi_-\mathcal{D}_{\mathrm{R}} + \text{ h.c.}\right] \,. \tag{3.5}$$

In this way, mixing between different quark families appears in the charged-current interaction, which is in accordance with observations.

3.2 CP Violation and Unitarity Triangles

In this section we shall discuss some properties of the the CKM matrix, in particular the CKM picture of CP violation.

From the above construction, it is clear that the (gauge) symmetries imply that the CKM matrix has to be unitary. A unitary $n \times n$ matrix has in general n^2 independent real parameters. However, in the case of the CKM matrix we may use our freedom to define the relative phases of the quark fields. For the case of n families we have n up-type and n down-type quarks, leaving us the freedom to chose $2n - 1$ relative phases. Consequently, the number of parameters N is

$$N = n^2 - 2n + 1 = (n-1)^2 \,. \tag{3.6}$$

Furthermore, if the CKM rotation were orthogonal (i.e. if the CKM matrix were real after we had used our freedom to rephase fields) it would have N_{angles} rotation angles; the remaining N_{phases} parameters necessarily would be phases. We obtain

$$N_{\text{angles}} = \frac{n(n-1)}{2}\,, \qquad N_{\text{phases}} = \frac{(n-1)(n-2)}{2}\,. \tag{3.7}$$

For the two-family case $n = 2$, there is only one parameter which is a rotation angle, the Cabibbo angle θ_C; furthermore, the CKM matrix is real and is an orthogonal 2×2 matrix. As we shall see below, this implies that a Standard Model with two generations cannot have CP violation, at least not for the minimal Higgs sector discussed in the previous chapter.

For the three-family case $n = 3$, we have four parameters and the CKM matrix may be written in terms of the sines and cosines of three angles and one complex phase factor. The case $n = 3$ is also the simplest case in which CP violation originating from the CKM matrix occurs and in the framework of the Standard Model this phase is in fact the only possible source of CP violation.

For $n = 3$, the CKM matrix may be understood as a product of three rotations in which one family always remains unchanged [2, 3, 4, 5, 6] This corresponds to the three Euler angles for a rotation in real, three-dimensional space. This leads us to define

$$U_{12} = \begin{bmatrix} c_{12} & s_{12} & 0 \\ -s_{12} & c_{12} & 0 \\ 0 & 0 & 1 \end{bmatrix}, \quad U_{13} = \begin{bmatrix} c_{13} & 0 & s_{13} \\ 0 & 1 & 0 \\ -s_{13} & 0 & c_{13} \end{bmatrix},$$
$$U_{23} = \begin{bmatrix} 1 & 0 & 0 \\ 0 & c_{23} & s_{23} \\ 0 & -s_{23} & c_{23} \end{bmatrix} \tag{3.8}$$

These three rotations define the three angles θ_{12}, θ_{13} and θ_{23}, where $c_{ij} = \cos\theta_{ij}$ and $s_{ij} = \sin\theta_{ij}$ are their cosines and sines.

The product of these three rotations yields a general orthogonal matrix, and if this were the CKM matrix, no CP violation would be possible. In order to obtain a CP-violating phase, we define another unitary matrix by

$$U_\delta = \begin{bmatrix} 1 & 0 & 0 \\ 0 & 1 & 0 \\ 0 & 0 & e^{-i\delta_{13}} \end{bmatrix}. \tag{3.9}$$

The standard parametrization of the CKM matrix, as proposed in [7] is given by a product of the three rotations, where U_{13} is transformed by the matrix U_δ:

$$V_{\text{CKM}} = U_{23} U_\delta^\dagger U_{13} U_\delta U_{12}\,. \tag{3.10}$$

Explicitly multiplying the matrices yields

$$V_{\text{CKM}} = \begin{pmatrix} c_{12}c_{23} & s_{12}c_{13} & s_{13}e^{-i\delta_{13}} \\ -s_{12}c_{13} - c_{12}s_{23}s_{13}e^{i\delta_{13}} & c_{12}c_{23} - s_{12}s_{23}s_{13}e^{i\delta_{13}} & s_{23}c_{13} \\ s_{12}s_{23} - c_{12}c_{23}s_{13}e^{i\delta_{13}} & -c_{12}s_{23} - s_{12}c_{23}s_{13}e^{i\delta_{13}} & c_{23}c_{13} \end{pmatrix}. \tag{3.11}$$

In the limit in which $\theta_{13} = \theta_{23} = 0$, the third generation decouples and the CKM matrix reduces to a orthogonal matrix describing Cabibbo mixing.

At present, the Particle Data Group [7] quotes the following range of values for the absolute values of the CKM matrix elements:

$$|V_{\text{CKM}}| = \begin{pmatrix} 0.9745 \text{ to } 0.9757 & 0.219 \text{ to } 0.224 & 0.002 \text{ to } 0.005 \\ 0.218 \text{ to } 0.224 & 0.9736 \text{ to } 0.9750 & 0.036 \text{ to } 0.046 \\ 0.004 \text{ to } 0.014 & 0.034 \text{ to } 0.046 & 0.9989 \text{ to } 0.9993 \end{pmatrix}, \quad (3.12)$$

where, for the last row, unitarity has been used. From these numbers it follows that $\theta_{12} \approx 12.7°$, $\theta_{23} \approx 2.3°$ and $\theta_{13} \approx 0.2°$, and hence $\theta_{12} \gg \theta_{23} \gg \theta_{13}$. This means that transitions within the same family are favoured. The further off the diagonal an entry is, the smaller is its absolute value. This is illustrated in Fig. 3.1.

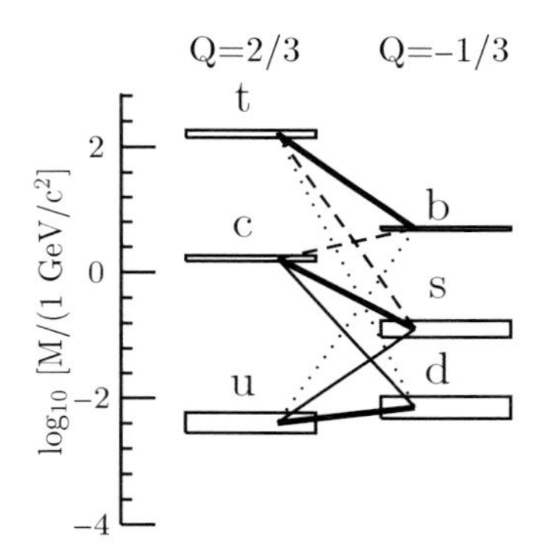

Fig. 3.1. Illustration of the relative strengths of charge current transitions

This phenomenological fact has led people to think of the CKM matrix in terms of an expansion in a small parameter λ [8], which can be chosen to be the sine of the Cabbibo angle $\lambda = |V_{us}|$. For the two-family case, we may write $\lambda = \sin\theta_C \approx \theta_C$, and obtain

$$V_{\text{CKM}} = \begin{pmatrix} 1 & 0 \\ 0 & 1 \end{pmatrix} + \begin{pmatrix} 0 & \lambda \\ -\lambda & 0 \end{pmatrix} + \begin{pmatrix} -\lambda^2/2 & 0 \\ 0 & -\lambda^2/2 \end{pmatrix} + \mathcal{O}(\lambda^3). \quad (3.13)$$

In the three-family case, we keep the same parameter λ and write

$$V_{\text{CKM}} = \begin{pmatrix} 1-\lambda^2/2 & \lambda & \lambda^3 A(\rho - i\eta(1-\lambda^2/2)) \\ -\lambda & 1-\lambda^2/2 - i\eta A^2\lambda^4 & \lambda^2 A(1+i\eta\lambda^2) \\ \lambda^3 A(1-\rho-i\eta) & -\lambda^2 A & 1 \end{pmatrix}, \quad (3.14)$$

where terms of order λ^4 in the real part and terms of order λ^5 in the imaginary part have been dropped. The three additional parameters A, ρ and η are all of order unity; from present data on B meson decays the values $A = 0.95 \pm 0.14$ and $\sqrt{\rho^2 + \eta^2} = 0.45 \pm 0.14$ are obtained [10].

Unitarity of the CKM matrix implies that the rows and columns of the matrix are orthonormal. In this way one can obtain 12 bilinear relations in total between the matrix elements. These are the six orthonormality conditions of the rows,

$$\sum_{q'=u,c,t} V_{q'q} V^*_{q'q''} = \delta_{qq''} \,, \tag{3.15}$$

and the six orthonormality conditions of the columns

$$\sum_{q'=d,s,b} V_{qq'} V^*_{q''q'} = \delta_{qq''} \,. \tag{3.16}$$

Since the matrix elements of the CKM matrix are in general complex-valued, these 12 relations may be depicted as triangles in the complex plane [9]. However, from consideration of the size of the CKM matrix elements, it is found that almost all of these triangles have one very small and two large sides, except for the two triangles

$$\sum_{q'=u,c,t} V^*_{q'b} V_{q'd} = V^*_{ub} V_{ud} + V^*_{cb} V_{cd} + V^*_{tb} V_{td} = 0 \,, \tag{3.17}$$

corresponding to the product of the first column with the complex conjugate of the last column, and

$$\sum_{q'=d,s,b} V^*_{uq'} V_{tq'} = V^*_{ud} V_{td} + V^*_{us} V_{ts} + V^*_{ub} V_{tb} = 0 \,, \tag{3.18}$$

corresponding to the product of the complex conjugate of the first row with the last row. These two triangles both have sides of order λ^3. However, owing to the unitarity of the CKM matrix they both correspond, up to terms of order λ^5, to the same relation between the Wolfenstein parameters ρ and η:

$$A\lambda^3(\rho + i\eta) - A\lambda^3 + A\lambda^3(1 - \rho - i\eta) = 0 \,. \tag{3.19}$$

The standard unitarity triangle is depicted in Fig. 3.2. In the Wolfenstein parametrization, it is a triangle in the ρ–η plane with a base of unit length, and its apex lies at the values of ρ and η given by (3.19).

The angles α, β and γ of the unitarity triangle are related to the phases of the CKM matrix elements. However, these angles are independent of the particular parametrization of the CKM matrix, and in the parametrization (3.11) one finds that, to leading order in the Wolfenstein parametrization, one has $\gamma = \delta_{13}$.

The non-vanishing phases in the CKM matrix imply CP violation in the Standard Model. In later applications we shall make us of the standard

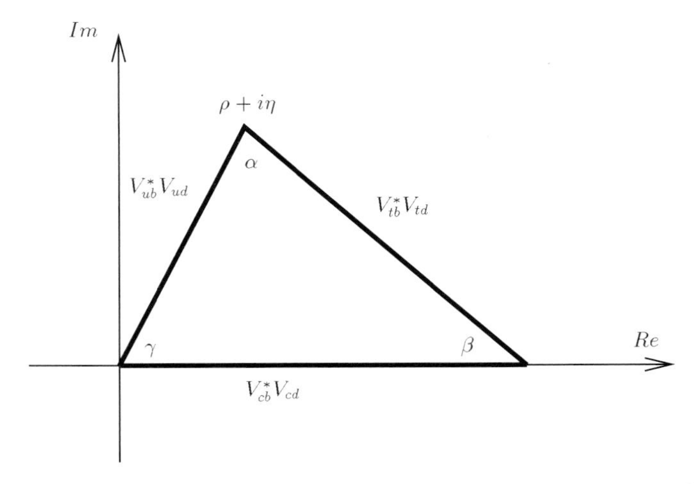

Fig. 3.2. The unitarity triangle, with the definition of the angles α, β and γ

parametrization, in which V_{ub} and V_{td} are the matrix elements that carry large phases, corresponding to the angles γ and β in Fig. 3.2. However, as can be seen from (3.11), the elements V_{cd}, V_{cs} and V_{ts} also carry phases, but these are tiny and do not appear in the Wolfenstein parametrization.

A non-vanishing phase $\delta_{13} \neq 0$ and $\delta_{13} \neq 180°$ means on the one hand a non-degenerate unitarity triangle, on the other hand it means that there is CP violation in the Standard Model. Since V_{CKM} is unitary, it can be shown that all 12 unitarity triangles have the same area and that this area is independent of the phase conventions used. Mathematically, this is related to the fourth-order rephasing invariants

$$\Delta_{\alpha\rho}^{(4)} = V_{\beta\sigma} V_{\gamma\tau} V_{\beta\tau}^* V_{\gamma\sigma}^* \,, \quad \text{where} \quad \begin{cases} \alpha, \beta, \gamma = u, c, t \text{ cyclic} \\ \rho, \sigma, \tau = d, s, b \text{ cyclic} \end{cases} , \tag{3.20}$$

and owing to the unitarity of the CKM matrix there is only one fourth-order rephasing invariant Δ. The imaginary part of Δ corresponds to the area of the unitarity triangles and hence may serve as a measure of CP violation [9]. Using the parametrization (3.11), one obtains

$$\operatorname{Im} \Delta = c_{12} s_{12} c_{13}^2 s_{13} s_{23} c_{23} \sin \delta_{13} \,, \tag{3.21}$$

which becomes simply $\operatorname{Im} \Delta = \lambda^6 A^2 \eta$ in the Wolfenstein parametrization.

In order to have non-vanishing CP violation, one has to have a non-zero $\operatorname{Im} \Delta$. This means that none of the angles θ_{ij} may take the values 0, 90° or 180°. On the other hand, $\operatorname{Im} \Delta$ has a maximal value of $1/(6\sqrt{3}) \approx 0.1$. This has to be compared with the value obtained from the measurement of CP violation in the kaon system where one finds $\operatorname{Im} \Delta \sim 10^{-4}$.

Finally, CP violation is also absent if any of the up- or down-type quarks are degenerate in mass. In this case one may perform a rotation among the two degenerate quarks which removes the CP-violating phase. It is, however, possible to define an invariant measure of CP violation by referring to the mass matrices defined in the previous section. It has been shown [10] that the determinant of the commutator of the two mass matrices

$$J = \det([\mathcal{M}_u\,, \mathcal{M}_d]) \tag{3.22}$$

is an invariant measure of CP violation, which is called the Jarlskog invariant. Explicit evaluation reveals that

$$\begin{aligned} J &= 2i\,\mathrm{Im}\,\Delta \\ &\times (m_u - m_c)(m_u - m_t)(m_c - m_t)(m_d - m_s)(m_d - m_b)(m_s - m_b)\,, \end{aligned} \tag{3.23}$$

showing explicitly that CP violation vanishes if mass degeneracies appear.

3.3 The CKM Matrix and the Fermion Mass Spectrum

In the Standard Model, the CKM matrix originates from the fact that the mass matrices (i.e. the matrices of Yukawa couplings) for the up and down quarks do not commute, and hence the mass eigenbasis for the up quarks is rotated relative to that for the down quarks, where the rotation is the CKM matrix. This strongly suggests a relation between the CKM matrix and the masses of the quarks. However, in the minimal version of the Standard Model the four parameters of the CKM matrix and the six quark masses are independent and thus unrelated parameters. The necessary information about the structure of the Yukawa matrices has to come from a theory beyond the Standard Model, which would need to explain the observed family structure. Without this information, one can only make assumptions about the matrices G and G' of Yukawa couplings, and these assumptions imply relations between the masses and the CKM matrix.

If we look at the quark mass spectrum, the only quark with a mass comparable to the weak scale (given by the vacuum expectation value of the Higgs field) is the top quark. Hence an ansatz where only the diagonal top-quark Yukawa coupling is non-vanishing can be used as a starting point. It was proposed some time ago [11] to use a rank-one matrix for the up quark, which may be cast into the form

$$G' \propto \begin{pmatrix} 1 & 1 & 1 \\ 1 & 1 & 1 \\ 1 & 1 & 1 \end{pmatrix} \tag{3.24}$$

by an appropriate rotation of the quark fields.

However, this ansatz does not yield a non-trivial CKM matrix, since the down-type quarks still all have vanishing mass and hence no mixing can occur. An ansatz for the down-quark mass matrix G has to have the property that it does not commute with G'. Thus, in a basis where G' is diagonal, G has to have non-zero off-diagonal entries.

There are a large number of ideas in the literature for inventing more or less justified matrices of Yukawa couplings, and we shall not even try to review these ideas; a recent review can be found in [12]. Rather, we restrict ourselves to a simple, text-book like example which at least shows a mechanism of how relations between masses and the CKM matrix may occur.

We restrict ourselves to two families, in which case we have four masses and one mixing angle. Our ansatz for the Yukawa couplings has to have fewer than five parameters in order to obtain the desired relations. Without restrictions, we may assume that the matrix of Yukawa couplings for the the up-type quarks is diagonal, where the diagonal entries are already two parameters, namely the masses of the up quarks. For the down type quarks we use the ansatz of a Hermitian matrix

$$G = \begin{pmatrix} 0 & a \\ a & 2b \end{pmatrix} , \tag{3.25}$$

where we shall assume that a and b are real and that the off-diagonal element a is much smaller than b. In this way, we have two more parameters a and b and thus we expect one relation between masses and mixing angles. Comparing this model with the general formulae of Sect. 3.1, we see that the unitary matrix diagonalizing G is already the CKM matrix of this simple toy model.

The two eigenvalues of the matrix G are

$$\lambda_1 = b + \sqrt{a^2 + b^2} \approx 2b \quad \text{and} \quad \lambda_2 = b - \sqrt{a^2 + b^2} \approx -\frac{a^2}{2b} , \tag{3.26}$$

and the CKM matrix in this toy model becomes

$$V_{\mathrm{CKM}} = \begin{pmatrix} \cos\theta & \sin\theta \\ -\sin\theta & \cos\theta \end{pmatrix} , \quad \text{where} \quad \tan\theta = \frac{a}{2b} . \tag{3.27}$$

On the other hand, the masses of the down-type quarks are the moduli of the eigenvalues of the matrix,

$$m_s = \lambda_1 v \approx 2bv \quad \text{and} \quad m_d = \lambda_2 v \approx \frac{a^2}{2b} v \tag{3.28}$$

where v is the vacuum expectation value of the Higgs field. Thus we end up with the relation

$$\tan\theta = \sqrt{\frac{m_d}{m_s}} . \tag{3.29}$$

Although this is only a toy model, the relation (3.29) is remarkably successful phenomenologically. Using the Particle Data Group range of the mass ratio $17 \le m_s/m_d \le 25$, we find values of the mixing angle between $11°$ and

14°; this has to be compared with the measured value of the Cabibbo angle $\theta_C \approx 12.7°$.

Despite its simplicity, this little toy model shows a general feature of many attempts to construct matrices of Yukawa couplings. The ansätze for these matrices have to contain fewer parameters than does the Standard Model, the parameters of which are the masses and the mixing angles. This is usually achieved by setting some of the matrix elements to zero, and there is a vast literature on deriving these "texture zeros" from e.g. symmetry considerations, for example [12].

The final answer to the question, of whether there is a relation between the CKM matrix and the mass spectrum and what it looks like has to wait for some more fundamental theory beyond the Standard Model. Moreover, the situation is different in the leptonic sector, since the possible right-handed neutrino does not carry any $SU(2)_L \times U(1)$ quantum number, and hence a Majorana mass term for these right handed neutrinos becomes possible, which is not generated by the Higgs mechanism. We shall make a few more remarks on this subject in the last chapter of this book.

References

1. S. Glashow, J. Iliopoulos and L. Maiani, Phys. Rev. D **2**, 1285 (1970).
2. M. Kobayashi and T. Maskawa, Prog. Theor. Phys. **49**, 652 (1973).
3. L. L. Chau and W. Y. Keung, Phys. Rev. Lett. **53**, 1802 (1984).
4. H. Harari and M. Leurer, Phys. Lett. B **181**, 123 (1986).
5. H. Fritzsch and J. Plankl, Phys. Rev. D **35**, 1732 (1987).
6. F. J. Botella and L. L. Chau, Phys. Lett. B **168**, 97 (1986).
7. K. Hagiwara et al. [Particle Data Group], Phys. Rev. D **66**, 010001 (2002).
8. L. Wolfenstein, Phys. Rev. Lett. **51**, 1945 (1983).
9. C. Jarlskog and R. Stora, Phys. Lett. B **208**, 268 (1988).
10. C. Jarlskog, Phys. Rev. Lett. **55**, 1039 (1985).
11. H. Fritzsch, Phys. Lett. B **70**, 436 (1977).
12. H. Fritzsch and Z. Xing, Prog. Part. Nucl. Phys. **45**, 1 (2000) [arXiv:hep-ph/9912358].

4 Effective Field Theories

4.1 What Are Effective Field Theories?

In describing a physical system, one can normally focus on the degrees of freedom that are relevant at the distance scales under consideration. For example, although it is well known that quantum mechanics is a more fundamental theory than classical mechanics, it would be difficult to describe the earth's motion around the sun by use of quantum mechanics. The state would correspond to a complicated superposition of energy eigenstates approximating the classical motion. Clearly, classical mechanics is the correct "effective theory".

In particle physics, the "correct" effective field theory is defined by distance or energy scales. Although we know that nuclei are composed of quarks, the appropriate degrees of freedom in nuclear physics are those of the nucleons, while the quark structure becomes relevant at much smaller distances. These smaller distance scales correspond to higher energies with which the system is probed.

In cases where very disparate mass scales appear, it is advantageous to construct an effective theory [1, 2, 3, 4, 5], where the degrees of freedom which become relevant only at much smaller distances (or, in other words, at much higher energy scales) do not appear explicitly. The most straightforward example is a heavy particle which cannot be created at an energy scale smaller than its mass; consequently, a Lagrangian valid at such small energies does not contain this degree of freedom. The fact that this is possible is ensured by the *decoupling theorem* proved by Applequist and Carazzone [6], who showed that – with very few exceptions – heavy degrees of freedom actually decouple at energy scales much lower than their mass. Decoupling means that any effect of these heavy degrees of freedom is (up to logarithmic contributions, which we shall discuss separately) suppressed by inverse powers of the heavy scale.

A case relevant to this book is that of weak interactions. All weak interactions among quarks are contained in the electroweak part of the Standard Model and in principle one could perform all calculations within the framework of the full Standard Model. However, when one considers decay processes of b hadrons (or even of lighter particles), the relevant scale of such a transition is the mass of the b quark, i.e. a scale of order $m_b \sim 5\,\mathrm{GeV}$, while the full Standard Model also contains very massive degrees of freedom (the top quark and the weak bosons, with masses of $\mathcal{O}(100 \text{ GeV})$), at least compared with the mass of the b quark. Thus it is advantageous to construct an effective theory from the full Standard Model in which the weak bosons and the heavy top quark do not appear explicitly any more [7]. We shall discuss this case in more detail in the next section.

The advantage of using an effective theory instead of the full theory is that many calculations simplify considerably. In particular, as we shall see below, using the renormalization group of the effective field theory allows us to perform resummations of large terms appearing in the radiative corrections.

The starting point for the construction of an effective field theory is the presence of a large scale Λ (usually the mass of a heavy particle), which in the case of weak interactions of hadrons is the mass of the weak boson M_W. The idea is to perform a separation of long- and short-distance contributions to transition matrix elements. Consider now some field theory (called the "full theory"), in which we consider a transition matrix element from some initial state $|i\rangle$ to a final state $|f\rangle$. In the case in which these states involve only energies $E_{i,f}$ lower than the heavy scale Λ, we can construct an effective Hamiltionian, since all effects of interactions from scales above Λ appear local at the typical scales of the states $|i\rangle$ and $|f\rangle$. In other words, the transition matrix elements for the interactions originating at the high scale Λ can be written as a matrix element of a local effective Hamiltonian $\mathcal{H}_{\text{eff}}$ [8],

$$\langle f|\mathcal{H}_{\text{eff}}|i\rangle = \sum_k C_k(\Lambda) \left. \langle f|\mathcal{O}_k|i\rangle \right|_\Lambda , \tag{4.1}$$

where $C_k(\Lambda)$ contains the short-distance contribution (i.e. the physics above the scale Λ), and the matrix elements $\langle f|\mathcal{O}_k|i\rangle|_\Lambda$ of the *local* operators $\mathcal{O}_k$ contain the long-distance contributions from scales below Λ.

The sum in (4.1) in general runs over an infinite set of operators, and hence (4.1) is only useful if we can truncate this infinite sum. The effective Hamiltonian is a density and thus has mass dimension four; hence the mass dimension of the short-distance coefficients $C_k(\Lambda)$ has to combine with the mass dimension of the operator in such a way that the total dimension of each term is four. Since the short-distance coefficients, by definition, do not depend on any long distance scale,[1] the mass dimension of the coefficients $C_k(\Lambda)$ has to come from powers of the large scale Λ. In order to simplify the counting of powers in $1/\Lambda$, it is convenient to factor out an appropriate power of $1/\Lambda$ and make the coefficient dimensionless. In this way the effective Hamiltonian can be written as

$$\langle f|\mathcal{H}_{\text{eff}}|i\rangle = \sum_k \frac{1}{\Lambda^k} \sum_i c_{k,i} \left. \langle f|\mathcal{O}_{k,i}|i\rangle \right|_\Lambda , \tag{4.2}$$

[1]This is of course connected to the fact that long and short distances can be factorized, which is non-trivial. Once factorization is proven, the mass dimension of the short-distance coefficients cannot originate from a long-distance scale.

where k is the dimension and we have taken into account the possibility that, for fixed dimension k, more than one operator (labelled by the subscript i) can contribute. In this normalization, the coefficients $c_{k,i}$ are dimensionless and hence – at least from naive dimensional arguments[2] – cannot depend on Λ [8].

The sum in (4.2) is thus ordered according to the dimension of the operators $\mathcal{O}_k$, and a truncation of the sum which neglects operators of mass dimension n corresponds to dropping terms of order $1/\Lambda^{n-4}$. Since the matrix elements contain only the long-distance scales of the states, their dimension is given by the energies of the states. In this way, one may construct a series expansion in powers of $E_{i,f}/\Lambda$. In the case of weak decays of hadrons this is a series in powers of m_{hadron}/M_W which converges rapidly.

In addition to these higher-dimensional operators, in general we still have dimension-four operators, which define a renormalizable theory, but in an effective theory operators of dimension larger than four appear in the way described above. However, these operators are not a problem concerning renormalization: the dimension-four terms of the effective action define a renormalizable theory, while all the higher-dimensional operators are suppressed by powers of the large scale, the inverse powers of which are used as an expansion parameter. Thus these higher-dimensional operators are inserted into the relevant Green's functions only as many times as are needed to compute to a definite order in the series in $1/\Lambda$, and, in a renormalizable theory, a finite number of insertions of higher-dimensional operators can always be renormalized. A detailed discussion of the subject of renormalization is beyond the scope of this book; a textbook presentation can be found in [9].

Before considering renormalization and the renormalization group, let us illustrate this idea with a simple example. If we consider the decay $b \to c\ell\bar{\nu}_\ell$, we can write the amplitude for this process in the full Standard Model. At tree level, this amplitude contains the propagator of a W boson between two left-handed currents. This process is depicted in the left Feynman diagram of Fig. 4.1. The maximal momentum transferred through this propagator is $q^2_{\text{max}} = (m_b - m_c)^2$, which is small compared with the W mass. Hence we can safely make the approximation

$$\frac{1}{M_W^2 - q^2} = \frac{1}{M_W^2}\left[1 + \frac{q^2}{M_W^2} + \cdots\right] , \tag{4.3}$$

which, in position space, corresponds to an expansion of the W propagator into local terms

---

[2]We shall see below that these naive arguments fail, since in a renormalizable theory the coupling constant, although dimensionless, depends on a dimensional quantity.

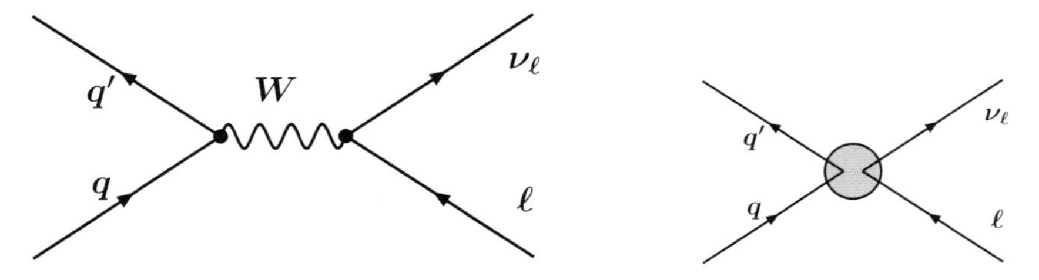

Fig. 4.1. Feynman diagram of the full theory (*left*) and of the effective theory (*right*). The *shaded dot* represents the insertion of the local effective Hamiltonian

$$\begin{aligned}\langle 0|T[W_\mu^+(x)W_\nu^-(0)]|0\rangle &= \int \frac{d^4q}{(2\pi)^4} e^{-iqx} \frac{(-i)g_{\mu\nu}}{M_W^2 - q^2} \\ &= \frac{(-i)g_{\mu\nu}}{M_W^2}\left[1 - \frac{\partial^2}{M_W^2} + \cdots\right]\delta^4(x) \ . \end{aligned} \tag{4.4}$$

This corresponds to the simple picture that the "range" of propagation of the W is $\mathcal{O}(1/M_W)$, which becomes local at distance scales of order $\mathcal{O}(1/m_b)$.

The transition amplitude corresponding to the first term may be written as a local effective Hamiltonian of the form

$$\mathcal{H}_{\text{eff}} = \frac{g^2}{8M_W^2}(\bar{b}\gamma_\mu(1-\gamma_5)c)(\bar{\nu}_\ell\gamma^\mu(1-\gamma_5)\ell) \ , \tag{4.5}$$

corresponding to the Feynman diagram on the right of Fig. 4.1. In this way one recovers the well-known Fermi interaction, which is indeed the leading term of a systematic expansion in inverse powers of the large weak-boson masses.

The separation of long and short distances does not require the presence of a degree of freedom with a heavy mass. One may equally well define an arbitrary scale parameter μ which has the dimensions of a mass, and all contributions to a matrix element above $\mu \leq \Lambda$ can be called short-distance pieces, while anything below μ belongs to the long-distance part. We may now apply the same arguments used previously for the scale Λ for the arbitrary scale μ, in which case (4.2) becomes

$$\langle f|\mathcal{H}_{\text{eff}}|i\rangle = \sum_k \frac{1}{\Lambda^k}\sum_i c_{k,i}(\Lambda/\mu)\ \langle f|\mathcal{O}_{k,i}|i\rangle\Big|_\mu \ , \tag{4.6}$$

where the (dimensionless) coefficients $c_{k,i}(\Lambda/\mu)$ contain the short-distance contribution (i.e. the physics above the scale μ), which may now depend on the ratio Λ/μ. The matrix elements $\langle f|\mathcal{O}_{k,i}|i\rangle|_\mu$ contain the long-distance contributions from scales below μ. In other words, changing μ moves contributions from the coefficient into the matrix element, and vice versa. The fact that no power corrections of order $1/\mu$ can appear is ensured by the renormalizability of the dimension-four part of the effective theory.

However, μ is an arbitrary scale parameter defining what are called the long- and short distance contributions. The requirement that the matrix elements are independent of μ is the origin of the renormalization group equations (a modern formulation can be found in [10]) frequently used in effective-field-theory calculations. These equations are derived from the requirement that a physical matrix element may not depend on this arbitrary scale μ. This has to hold for the matrix element of the effective Hamiltonian, and thus we obtain

$$0 = \mu \frac{d}{d\mu} \langle f | \mathcal{H}_{\text{eff}} | i \rangle \, . \tag{4.7}$$

The effect of a change in μ is twofold. First of all, lowering μ shifts parts of the matrix elements of the operators $\mathcal{O}_{k,i}$ into the short-distance coefficients $C_{k,i}$; secondly, the operators "mix" under renormalization; this means that, starting at some scale μ, the contribution from a matrix element of an operator $\mathcal{O}_{k,i}$ turns at some other scale μ' into a sum of contributions from all the matrix elements of the operators $\mathcal{O}_{k,j}$ which have the same mass dimension and quantum numbers as the original operator $\mathcal{O}_{k,i}$.[3]

In order to derive these equations, we perform the differentiation in (4.7) using (4.6). The relations that we are going to obtain will hold for each power of Λ separately, and to keep things simple we shall drop the index k labelling the dimension of the operator. Thus, in the following, the operators $\mathcal{O}_i$ are a set of operators which have the same dimension; in fact they have to form a basis closed under renormalization, i.e. any operator generated by mixing can be written as a linear combination of the basis operators. We find

$$0 = \sum_i \left[\left(\mu \frac{d}{d\mu} c_i(\Lambda/\mu) \right) \langle f | \mathcal{O}_i | i \rangle \Big|_\mu + c_i(\Lambda/\mu) \left(\mu \frac{d}{d\mu} \langle f | \mathcal{O}_i | i \rangle \Big|_\mu \right) \right] \, . \tag{4.8}$$

Mixing implies that, under an infinitesimal change in $\log \mu$, the operator $\mathcal{O}_i$ turns into a linear combination of operators with the same mass dimension:

$$\mu \frac{d}{d\mu} \langle f | \mathcal{O}_i | i \rangle \Big|_\mu = \sum_j \gamma_{ij}(\mu) \langle f | \mathcal{O}_j | i \rangle \Big|_\mu \, , \tag{4.9}$$

where the matrix γ is called the anomalous-dimension matrix, and depends on the scale μ. Inserting this, we obtain

$$\begin{aligned} 0 &= \sum_i \left(\mu \frac{d}{d\mu} c_i(\Lambda/\mu) \right) \langle f | \mathcal{O}_i | i \rangle \Big|_\mu + \sum_i \sum_j c_i(\Lambda/\mu) \gamma_{ij}(\mu) \langle f | \mathcal{O}_j | i \rangle \Big|_\mu \\ &= \sum_i \sum_j \left[\delta_{ij} \mu \frac{d}{d\mu} + \gamma_{ij}(\mu) \right] c_i(\Lambda/\mu) \langle f | \mathcal{O}_j | i \rangle \Big|_\mu \, . \end{aligned} \tag{4.10}$$

[3] The fact that mixing occurs only with operators of the same dimension (or at worst with lower-dimensional operators) is again due to renormalizability [9]. We shall also assume that a mass-independent regularization is used such that operator mixing appears only among operators of the same mass dimension.

As stated above, the operators $\mathcal{O}_i$ form a basis for the operators of a fixed dimension, which means that none of the operators may be written as a linear combination of the others. Consequently (4.10) is equivalent to the set of equations for the coefficients

$$\sum_i \left[\delta_{ij} \mu \frac{d}{d\mu} + \gamma_{ij}^T(\mu) \right] c_j(\Lambda/\mu) = 0 \,. \tag{4.11}$$

Note that, owing to the definition of the anomalous-dimension matrix (4.9), the transpose of this matrix appears in the renormalization group (4.11) for the Wilson coefficients.

The anomalous dimension is a dimensionless quantity and, from a naive point of view, cannot depend on the mass scale μ. However, in a renormalizable theory there is a "hidden" scale, which is given by the scale dependence of the coupling constant [9]. In other words, in a renormalizable theory such as QCD one may arrange things in such a way that, for observable quantities, a change in scale may be compensated by an appropriate change in the masses and the coupling constants of the theory.

In the following we shall consider the case of massless QCD, which is the case considered throughout this book. In other words, we shall consider only the renormalization group flow induced by strong interactions. In the case of massless QCD only the strong coupling constant changes with scale. This change is governed by the equation for the running coupling

$$\mu \frac{d}{d\mu} \alpha_s(\mu) = \beta(\alpha_s(\mu)) \,, \tag{4.12}$$

where the function $\beta(\alpha_s)$ depends on μ only through α_s, which is used as a parameter for a perturbative expansion of the β function.

Similarly, the anomalous-dimension matrix $\gamma_{kj}(\mu)$ depends on μ only through the μ dependence of the strong coupling constant α_s, i.e.

$$\gamma_{ij}(\mu) = \gamma_{ij}(\alpha_s(\mu)) \,. \tag{4.13}$$

Like the β function, the anomalous-dimension matrix will be expanded in powers of the strong coupling.

The coefficients c_i depend on μ not only through Λ/μ but also through their dependence on the coupling α_s, since the coupling is scale-dependent in QCD. This means that the total derivative with respect to μ must be replaced by

$$\mu \frac{d}{d\mu} = \left(\mu \frac{\partial}{\partial \mu} + \beta(\alpha_s) \frac{\partial}{\partial \alpha_s} \right) \,. \tag{4.14}$$

Inserting this, we may rewrite the renormalization group equation as

$$\sum_i \left[\delta_{ij} \left(\mu \frac{\partial}{\partial \mu} + \beta(\alpha_s) \frac{\partial}{\partial \alpha_s} \right) + \gamma_{ij}^T(\alpha_s) \right] c_j(\Lambda/\mu, \alpha_s) = 0 \,, \tag{4.15}$$

where we display explicitly the α_s dependence of the coefficients c_i.

The renormalization group equation is a linear partial differential equation which, for known β and γ_{jk} functions, has unique solutions once initial values of the coefficients at some scale μ_0 are given. The original idea for constructing an effective Hamiltonian was to move the effects of scales above Λ into the short-distance coefficients c_i appearing in (4.2). This means that the coefficients c_i are determined by comparing the full theory with the effective theory at the scale Λ, which is usually called matching. Thus the c_i in (4.2) are actually the coefficients at $\mu = \Lambda$, i.e. $c_i(\Lambda/\mu = 1, \alpha_s(\Lambda))$, and thus the matching calculation yields the initial condition for the renormalization group flow of the coefficients c_i.

In any practical application, the coefficients $c_{k,i}$ as well as the anomalous dimensions and the β function, are computed in perturbation theory as a series in α_s. Thus the coefficients take the general form

$$c_i(\Lambda/\mu = 1, \alpha_s) = \sum_n a_i^{(n)} \left(\frac{\alpha_s}{4\pi}\right)^n , \tag{4.16}$$

and the β function and the anomalous dimensions have the form

$$\beta(\alpha_s) = \alpha_s \sum_{n=0} \beta^{(n)} \left(\frac{\alpha_s}{4\pi}\right)^{n+1} , \qquad \gamma_{ij}(\alpha_s) = \sum_{n=0} \gamma_{ij}^{(n)} \left(\frac{\alpha_s}{4\pi}\right)^{n+1} \tag{4.17}$$

where we have taken into account the fact that the first non-vanishing terms in the β function are of second order, and (usually) of first order in the anomalous dimensions. For later use, we quote the well-known first non-vanishing term of the β function:

$$\beta^{(0)} = -\frac{2}{3}(33 - 2n_{\mathrm{f}}) , \tag{4.18}$$

where n_{f} is the number of active flavours, i.e. the number of quarks with masses below the scale μ. We shall not go into any more details concerning renormalization and computation of the perturbative series for the β function and the anomalous dimensions; for this, we refer the reader to textbooks such as [9] or the review paper [11].

Taking the perturbative expansion as an input for the renormalization group equations, we may compute the coefficients c_i at some lower scale μ. Expanded in a perturbative series, the coefficients become

$$\begin{aligned} c_i(\Lambda/\mu, \alpha_s) = {} & b_i^{00} \\ & + b_i^{11} \left(\frac{\alpha_s}{4\pi}\right) \ln\frac{\Lambda}{\mu} + b_i^{10} \left(\frac{\alpha_s}{4\pi}\right) \\ & + b_i^{22} \left(\frac{\alpha_s}{4\pi}\right)^2 \ln^2\frac{\Lambda}{\mu} + b_i^{21} \left(\frac{\alpha_s}{4\pi}\right)^2 \ln\frac{\Lambda}{\mu} + b_i^{20} \left(\frac{\alpha_s}{4\pi}\right)^2 \\ & + b_i^{33} \left(\frac{\alpha_s}{4\pi}\right)^3 \ln^3\frac{\Lambda}{\mu} + b_i^{32} \left(\frac{\alpha_s}{4\pi}\right)^3 \ln^2\frac{\Lambda}{\mu} + b_i^{31} \left(\frac{\alpha_s}{4\pi}\right)^3 \ln\frac{\Lambda}{\mu} + \cdots , \end{aligned} \tag{4.19}$$

where the superscripts of the coefficients b_i denote the power of α_s and the power of the logarithm $\ln(\Lambda/\mu)$. In particular, at $\Lambda = \mu$ all the logarithms vanish and we have $b_i^{(n0)} = a_i^{(n)}$.

However, when we know the perturbative expansion of the renormalization group functions β and γ_{ij}, the renormalization group equations allow us to resum the columns of the perturbative result (4.19). If we use the first non-vanishing terms $\beta^{(0)}$ and $\gamma_{ij}^{(0)}$, the solution of the renormalization-group equation contains all orders of α_s and performs a resummation of all contributions with coefficients $b_{k,i}^{nn}$, where the power of the logarithm is equal to the power of α_s. This is called the *leading-logarithmic approximation* (LLA). For the case where only a single operator with dimension k appears (i.e. no mixing can occur), we can solve the renormalization group equation to obtain

$$c_i(\Lambda/\mu, \alpha_s(\Lambda)) = b_i^{00} \left(\frac{\alpha_s(\Lambda)}{\alpha_s(\mu)} \right)^{\gamma^{(0)}/\beta^{(0)}} . \tag{4.20}$$

Note that in this case the matching calculation needs to be performed only at tree level, i.e. only the coefficient b_i^{00} is needed. Taking as the expansion parameter, for example, $\alpha_s(\Lambda)$ one immediately reproduces the first column of (4.19) in terms of $\beta^{(0)}$ and $\gamma^{(0)}$

Going beyond the LLA requires us to know the next term in the perturbative expansion of the renormalization group functions β and γ_{ij}. If, in addition to this, the next term of the expansion of c_i (which is the coefficient $b_{k,i}^{10}$) is known, one can resum the second column of (4.19), i.e. one can sum the terms involving the coefficients $b_i^{n\,n-1}$. However, this requires a matching calculation at order α_s which is usually a (complete) one-loop calculation.

Using the renormalization group machinery thus allows us to move contributions from the matrix element to the coefficients c_i, thereby resumming logarithms $\ln(\Lambda/\mu)$ in a systematic way. Ideally, one would like μ to be a typical scale appearing in the matrix element, such as the mass of the decaying hadron. On the other hand, since we are using perturbation theory for the calculation of the coefficients c_i, μ has to be a perturbative scale. For applications to weak interactions Λ is of the order of the weak-boson mass M_W, while it is usually assumed that the charm quark mass $m_c \sim 1.5\,\text{GeV}$ is still a perturbative scale. Taking this as an estimate we see that $\ln(\Lambda/\mu) = \ln(M_W/m_c) \sim 4$ is relatively large and that the contribution of the term $b_i^{22}(\alpha_s/\pi)^2 \ln^2(\Lambda/\mu)$ could overwhelm that of $b_i^{10}(\alpha_s/\pi)$ even though it is formally of higher order in perturbation theory. In such cases a resummation becomes mandatory, and it is the strength of the effective-field-theory approach that this resummation can be performed using the renormalization group.

4.2 Fermi's Theory as an Effective Field Theory

In this section we shall review Fermi's theory in the light of the above discussion of effective theories. It turns out that Fermi's theory is an effective theory which, on one hand, can be obtained from the Standard Model in the limit of infinitely heavy weak gauge bosons (and infinitely heavy top quark) [7], and, on the other hand, is the most general effective theory that exhibits a *local* $SU(3) \times SU(2) \times U(1)$ symmetry [12]. The interesting (and perhaps not so well known) fact is that (except for $SU(3) \times U(1)_{em}$, which remains unbroken) the electroweak symmetry is implemented as a local symmetry *without gauge fields*. However, this happens at the cost that dimension-six operators appear and hence the resulting Lagrangian has to be interpreted as an effective field theory.

In order to present the argument we shall start with a simplified version with only a single doublet of left handed quarks and ignore the mass term. Furthermore, we shall discuss only the left-handed $SU(2)$ symmetry, ignoring for the moment the presence of the $U(1)$. Formally, this corresponds to the case $\Theta_W \to 0$ and vanishing quark masses. The Lagrangian of this simplified version is

$$\mathcal{L}_{kin} = \bar{Q} i \partial\!\!\!/ Q + \frac{v^2}{4} \mathrm{Tr} \left[(\partial_\mu \Sigma)^\dagger (\partial^\mu \Sigma) \right] , \tag{4.21}$$

where we shall make use of the non-linear representation of the Higgs field where $H \to v\Sigma$ and $\Sigma\Sigma^\dagger = \Sigma^\dagger\Sigma = 1$ (see Sect. 2.2). This Lagrangian has a global $SU(2)_L$ symmetry with

$$\Sigma \to \Lambda\Sigma , \quad Q \to \Lambda Q , \quad \Lambda \in SU(2)_L . \tag{4.22}$$

Note that Σ has a non-vanishing vacuum expectation value and hence $SU(2)_L$ is spontaneously broken. In fact, the three degrees of freedom in Σ (see (2.24)) are the Goldstone modes of this symmetry breaking. These massless modes are not present in nature, but their appearance can be avoided by promoting the global $SU(2)_L$ symmetry to a *local* symmetry. This requires us to introduce gauge fields, which are used to construct a covariant derivative. We shall not go into any detail on gauge theories here, and refer the reader to textbooks on this subject [13, 14, 15, 16, 17, 18, 19].

However, for the case at hand, we have to introduce three gauge fields W^a_μ, $a = 1, 2, 3$ (corresponding to the charged W bosons and the neutral Z), but these will only be auxiliary fields. We introduce the covariant derivative in the usual way:

$$\partial_\mu Q \to D_\mu Q = \partial_\mu Q - \frac{i}{2} g W^a_\mu \tau^a Q , \tag{4.23}$$

$$\partial_\mu \Sigma \to D_\mu \Sigma = \partial_\mu \Sigma - \frac{i}{2} g W^a_\mu \tau^a \Sigma , \tag{4.24}$$

$$\partial_\mu \Sigma^\dagger \to (D_\mu \Sigma)^\dagger = \partial_\mu \Sigma^\dagger - \frac{i}{2} g W^a_\mu \Sigma^\dagger \tau^a , \tag{4.25}$$

where the τ^a are the usual Pauli matrices and the gauge fields have the proper behaviour under $SU(2)_L$ gauge transformations

The usual way to construct a gauge theory is to introduce kinetic-energy terms for all gauge fields. However, here we treat the W^a only as auxiliary fields, which means that we omit their kinetic-energy contribution from the Lagrangian. Owing to the transformation properties of the gauge fields, the Lagrangian

$$\mathcal{L} = \bar{Q} i \not{D} Q + \frac{v^2}{4} \mathrm{Tr}\left[(D_\mu \Sigma)^\dagger (D^\mu \Sigma) \right] \tag{4.26}$$

is now invariant under *local* $SU(2)_L$ transformations.

As the next step, we can write all terms of the Lagrangian explicitly, which gives

$$\begin{aligned} \mathcal{L} &= \bar{Q} i \not{\partial} Q + \frac{v^2}{4} \mathrm{Tr}\left[(\partial_\mu \Sigma)^\dagger (\partial^\mu \Sigma) \right] \\ &+ \frac{v^2}{4} g W_\mu^a \left[J^{\mu,a} + \frac{v^2}{4} j^{\mu,a} \right] + \frac{v^2}{8} g^2 W_\mu^a W^{\mu,a} , \end{aligned} \tag{4.27}$$

where we have defined

$$J_\mu^a = \frac{i}{2} \mathrm{Tr}\left\{ \Sigma^\dagger \tau^a (\partial_\mu \Sigma) - (\partial_\mu \Sigma^\dagger) \tau^a \Sigma \right\} , \tag{4.28}$$

$$j_\mu^a = \frac{1}{2} \bar{Q} \gamma_\mu \tau^a Q . \tag{4.29}$$

As stated above, the fields W_μ^a are only auxiliary degrees of freedom, which means that they have an algebraic equation of motion. Replacing these fields using this equation of motion (or, alternatively, removing all interactions by quadrature and integrating out the auxiliary fields), we obtain

$$\begin{aligned} \mathcal{L} &= \bar{Q} i \not{\partial} Q + \frac{v^2}{4} \mathrm{Tr}\left[(\partial_\mu \Sigma)^\dagger (\partial^\mu \Sigma) \right] \\ &- \frac{v^2}{8} \left[J^{\mu,a} + \frac{v^2}{4} j^{\mu,a} \right] \left[J_\mu^a + \frac{v^2}{4} j_\mu^a \right] , \end{aligned} \tag{4.30}$$

which still is *locally* $SU(2)_L$ invariant.[4] In fact, it is an easy exercise to show that the contribution appearing, for example, from the non-invariance of the kinetic term of fermions is compensated by a term that originates from the product of the currents in the second line of (4.30).

Making use of this local symmetry, we may choose to use the unitary gauge, where all Goldstone modes are gauged away, in which case we have $\Sigma \equiv 1$ and thus $J_\mu^a \equiv 0$, and the Lagrangian becomes

$$\mathcal{L} = \bar{Q} i \not{\partial} Q - \frac{1}{2v^2} \left(\bar{Q} \gamma_\mu \tau^a Q \right) \left(\bar{Q} \gamma^\mu \tau^a Q \right) , \tag{4.31}$$

[4] $\mathcal{L}$ is not renormalizable any more, but we are considering it only as an effective field theory valid at scales much less than v.

which is the Fermi interaction in the limit $\Theta_W = 0$. Thus one can regard this interaction as the result obtained in the unitary gauge of a spontaneously broken *local* $SU(2)_L$ symmetry.

Of course, we can derive the same result for the full $SU(2)_L \times U(1)_Y$ symmetry of the Standrd Model. We start again from an $SU(2)_L \times SU(2)_R$ symmetry and use the same notation as in the last chapter. in addition to quarks we introduce Higgs fields using the non-linear representation, such that the kinetic term becomes (see (2.17))

$$\mathcal{L}_{kin} = \sum_i \left[\bar{Q}_i \not{\partial} Q_i + \bar{q}_i \not{\partial} q_i\right] + \frac{v^2}{4}\mathrm{Tr}\left[(\partial_\mu \Sigma)^\dagger(\partial^\mu \Sigma)\right] . \tag{4.32}$$

As before, we can write $SU(2) \times SU(2)$-invariant (custodial-symmetry-conserving) and $SU(2)\times U(1)$-invariant (custodial-symmetry-breaking) terms, yielding

$$\mathcal{L} = \mathcal{L}_{kin} - v\sum_{ij} y'_{ij}\bar{Q}_i \Sigma q_j - v\sum_{ij} y'_{ij}\bar{Q}_i \Sigma T_{3,R} q_j + \text{ h.c.} , \tag{4.33}$$

which contains the mass term since $\langle 0|\Sigma|0\rangle = 1$.

The Lagrangian (4.33) has a global $SU(2) \times U(1)$ symmetry, which is spontaneously broken down to $U(1)_{em}$. As in the simpler example, we can again remove the massless Goldstone modes by gauging (4.33), i.e. by promoting the global $SU(2) \times U(1)$ symmetry to a *local* one. The gauge fields that are needed are the W, the Z boson and the A field. Since the electromagnetic $U(1)$ symmetry remains unbroken, we can keep the corresponding covariant derivative and write

$$\begin{aligned}\partial_\mu Q_A \to D_\mu Q_A = \mathcal{D}_\mu Q_A - i\frac{g}{\sqrt{2}}\left(W_\mu^+\tau^+ + W_\mu^-\tau^-\right)Q_A \\ -i\frac{g}{2\cos\theta_W}Z_\mu\left[\cos^2\theta_W\tau^3 - \sin^2\theta_W\frac{1}{3}\right]Q_A ,\end{aligned} \tag{4.34}$$

$$\partial_\mu q_A \to D_\mu q_A = \mathcal{D}_\mu q_A - i\frac{g}{2\cos\theta_W}\sin^2\theta_W Z_\mu\left[\tau^3 + \frac{1}{3}\right]q_A , \tag{4.35}$$

$$\begin{aligned}\partial_\mu \Sigma \to D_\mu \Sigma = \mathcal{D}_\mu \Sigma - i\frac{g}{\sqrt{2}}\left(W_\mu^+\tau^+ + W_\mu^-\tau^-\right)\Sigma \\ -i\frac{g}{2\cos\theta_W}Z_\mu\left[\cos^2\theta_W\tau^3\Sigma - \sin^2\theta_W\Sigma\tau^3\right] ,\end{aligned} \tag{4.36}$$

where $\tau^\pm$ and τ^3 are the usual Pauli matrices and $\mathcal{D}$ denotes the covariant derivative with respect to the electromagnetic $U(1)_{em}$ symmetry (and also with respect to the colour $SU(3)$ symmetry, which we do not consider here).

We can again remove the auxiliary fields $W^\pm$ and Z, since these fields have an algebraic equation of motion. However, no mass term appears for the electromagnetic field, which becomes a dynamic degree of freedom after a kinetic-energy term is included. We shall not discuss the electromagnetic contribution any more, and set $A_\mu \equiv 0$ in the following.

After eliminating the $W^\pm$ and Z we may again go to the unitary gauge, in which case all contributions from the Goldstone modes vanish. The Lagrangian in the unitary gauge becomes

$$\begin{aligned}
\mathcal{L} = &\sum_i \left[\bar{Q}_i \not{\partial} Q_i + \bar{q}_i \not{\partial} q_i\right] \\
&- \left[v \sum_{ij} y'_{ij} \bar{Q}_i q_j + v \sum_{ij} y'_{ij} \bar{Q}_i T_{3,R} q_j + \text{ h.c.} \right] \\
&- \frac{G_F}{\sqrt{2}} \left[\bar{Q}_i \tau^+ \gamma_\mu Q_i\right] \left[\bar{Q}_j \tau^- \gamma_\mu Q_j\right] \\
&- \frac{G_F}{2\sqrt{2}} \left[\bar{Q}_i \left(\cos^2\theta_W \tau^3 - \frac{1}{3}\cos^2\theta_W \right) \gamma_\mu Q_i + \sin^2\theta_W \bar{q}_i \left(\tau^3 + \frac{1}{3} \right) \gamma_\mu q_i \right] \\
&\times \left[\bar{Q}_i \left(\cos^2\theta_W \tau^3 - \frac{1}{3}\cos^2\theta_W \right) \gamma^\mu Q_i + \sin^2\theta_W \bar{q}_i \left(\tau^3 + \frac{1}{3} \right) \gamma^\mu q_i \right] .
\end{aligned} \tag{4.37}$$

The interesting point about the Lagrangian (4.37) is that it is the unitary-gauge version of a Lagrangian which is invariant under *local* $SU(2) \times U(1)$ transformations, although the gauge fields do not appear any more. However, the price to be paid is that (4.37) is not renormalizable and must be interpreted as an effective Lagrangian, valid for energies much lower than the weak scale v.

Custodial $SU(2)$ symmetry is also present in the Lagrangian (4.37). In the limit $\theta_W = 0$, the three components of the weak current (i.e. the two charged currents and the neutral current) form a triplet under custodial $SU(2)$, and the symmetry of the Lagrangian implies that the coupling strengths of the charged and neutral currents are the same in this limit. This is the low-energy manifestation of custodial $SU(2)$. The breaking of custodial $SU(2)$ occurs explicitly through a non-vanishing value of θ_W and the Yukawa couplings, more precisely the second term in the second line of (4.37).

Diagonalizing the Yukawa-coupling matrices corresponds to the mass eigenbasis for the quarks. In the same way as in Sect. 2.2 this affects only the charged currents, which means that the CKM matrix appears in the third line of (4.37), rotating the down-type quarks by the usual CKM rotation.

The Lagrangian (4.37) is the Lagrangian of Fermi's theory of weak interactions, which we have derived from the assumption of a spontaneously broken, local $SU(2) \times U(1)$ symmetry. Clearly it is also the limit of the Standard Model for large gauge boson masses, since in this limit the kinetic-energy term for the gauge bosons becomes irrelevant compared with their mass term, which, owing to the Higgs mechanism, is contained in the kinetic-energy term of the scalar particles.

4.3 Heavy-Quark Effective Theory

Another example of an effective field theory is the effective theory for heavy quarks (heavy-quark effective theory, HQET).[5] This effective theory describes a quark with a mass m_Q much larger than Λ_{QCD}, the scale parameter of QCD. In this case the mass of the heavy quark is still a perturbative scale such that $\alpha_s(m_Q)$ is small enough to allow a perturbative treatment.

Unlike in the Fermi theory of weak interactions, where the heavy particles (the top quark and the W boson) do not appear any more in the effective field theory, in HQET there is still a remnant of the heavy quark at scales below m_Q. In QCD, flavour numbers are conserved, and hence a heavy quark with a definite flavour cannot decay and thus is present at any scale. What remains at scales much less than m_Q is a static source of colour [30], which acts very similarly to the heavy proton inside a hydrogen atom.

This heavy-mass limit, which has been known since the 1930s [31], can be formulated as an effective theory. In addition, one can construct the effective Hamiltonian explicitly by integrating out heavy degrees of freedom from the functional integral of QCD Green's functions. This integration may be performed explicitly [32], since in the case at hand it amounts to a Gaussian functional integration. We start from the generating functional of the QCD Green's functions

$$
\begin{aligned}
&Z(\eta, \bar{\eta}, \lambda) \\
&= \int [dQ][d\bar{Q}][d\phi_\lambda] \exp\left\{ iS + iS_\lambda + i \int d^4x \, (\bar{\eta} Q + \bar{Q}\eta + \phi_\lambda \lambda) \right\}, \quad (4.38)
\end{aligned}
$$

where $\phi_\lambda = q$, A^a_μ denotes the light degrees of freedom (light quarks q and gluons A_μ) with an action S_λ, while S denotes the piece of the action for the heavy quark Q, including its coupling to the gluons,

$$
S = \int d^4x \, \bar{Q}(i \not{D} - m_Q) Q \, , \quad (4.39)
$$

where

$$
D_\mu = \partial_\mu + igA_\mu \quad (4.40)
$$

is the covariant derivative of QCD. We have introduced source terms η for the heavy quark and λ for the light degrees of freedom.

We shall consider hadrons containing a single heavy quark, and we assume that this heavy hadron moves with a certain velocity v,

$$
v = \frac{p_{\text{hadron}}}{m_{\text{hadron}}}; \quad v^2 = 1, \quad v_0 > 0 \, . \quad (4.41)
$$

[5] The original articles on this subject are [20, 21, 22, 23, 24, 25], review articles are [26, 27, 28], and a textbook presentation is given in [29].

The way to proceed is along the lines of the non-relativistic reduction of the Dirac equation: the velocity vector may be used to split the heavy-quark field Q into an "upper" component ϕ and a "lower" component χ

$$\phi_v = \frac{1}{2}(1 + \not{v})Q, \ \ \not{v}\phi_v = \phi, \tag{4.42}$$

$$\chi_v = \frac{1}{2}(1 - \not{v})Q, \ \ \not{v}\chi_v = -\chi, \tag{4.43}$$

and to define a decomposition of the covariant derivative into a "longitudinal" and a "transverse" ($\perp$) part

$$D_\mu = v_\mu(v \cdot D) + D^\perp_\mu \, , \quad D^\perp_\mu = (g_{\mu\nu} - v_\mu v_\nu)D^\nu, \quad \{\not{D}^\perp \, , \not{v}\} = 0. \tag{4.44}$$

Using (4.42–4.44) the action (4.39) of the heavy quark field takes the form

$$S = \int d^4x \left[\bar{\phi}\{i(v\cdot D) - m_Q\}\phi - \bar{\chi}\{i(v\cdot D) + m_Q\}\chi + \bar{\phi} i\not{D}^\perp\chi + \bar{\chi} i\not{D}^\perp\phi\right]. \tag{4.45}$$

The heavy quark in a meson is very close to being on shell, and thus the space–time dependence of the heavy-quark field is mainly that of a free particle moving with velocity v. This suggests a reparametrization of the fields by removing the space–time dependence of a solution of the free Dirac equation. We shall chose the "particle-type" parametrization, where we pick out the "positive-energy solution" of the Dirac equation

$$\phi_v = e^{-im_Q(v\cdot x)}h_v \, , \qquad \chi_v = e^{-im_Q(v\cdot x)}H_v, \tag{4.46}$$

such that the space time dependence of the remaining fields h_v and H_v is determined by the residual momentum $k = p - m_Q v$, which is due to binding effects of the heavy quark inside the heavy hadron and is a "small" quantity of order Λ_{QCD}.

Expressed in these fields, the action of the heavy quark becomes

$$S = \int d^4x \times \left[\bar{h}_v i(v \cdot D)h_v - \bar{H}_v\{i(v \cdot D) + 2m_Q\}H_v + \bar{h}_v i\not{D}^\perp H_v + \bar{H}_v i\not{D}^\perp h_v\right] \, . \tag{4.47}$$

The term containing the sources is also rewritten in terms of the fields h_v and H_v:

$$\int d^4x \, (\bar{\eta}\psi + \bar{\psi}\eta) = \int d^4x \, (\bar{\rho}_v h_v + \bar{h}_v\rho_v + \bar{R}_v H_v + \bar{H}_v R_v), \tag{4.48}$$

where ρ_v and R_v are now source terms for the upper-component field h_v and the lower component part H_v, respectively.

In terms of the new variables, the generating functional reads

$$Z(\rho_v, \bar{\rho}_v, R_v, \bar{R}_v, \lambda) = \int [dh_v][d\bar{h}_v][dH_v][d\bar{H}_v][d\phi_\lambda]$$
$$\times \exp\left\{ iS + S_\lambda + i \int d^4x \, (\bar{\rho}_v h_v + \bar{h}_v \rho_v + \bar{R}_v H_v + \bar{H}_v R_v + \phi_\lambda \lambda) \right\} , \tag{4.49}$$

where the action S for the heavy quark is given in (4.47).

From (4.47) it is obvious that the heavy degree of freedom is the lower-component field H_v, since it has a mass term $2m_Q$, while the upper component field h_v is a massless field describing the static heavy quark. In the heavy-mass limit only the Green's functions involving the field h_v have to be calculated, and hence we integrate over H_v in the functional integral (4.49) with the sources of the lower-component field R_v and $\bar{R}_v$ set to zero. This can be done explicitly, since it is a Gaussian integration

$$Z(\rho_v, \bar{\rho}_v, \lambda) = \int [dh_v][d\bar{h}_v][d\lambda] \, \Delta$$
$$\times \exp\left\{ iS + S_\lambda + i \int d^4x \, (\bar{\rho}_v^+ h_v^+ + \bar{h}_v^+ \rho_v^+ + \phi_\lambda \lambda) \right\} , \tag{4.50}$$

where now the action functional for the heavy quark becomes a non-local object

$$S = \int d^4x \left[\bar{h}_v^+ i(v \cdot D) h_v^+ - \bar{h}_v^+ \not{D}^\perp \left(\frac{1}{i(v \cdot D) + 2m_Q - i\epsilon} \right) \not{D}^\perp h_v^+ \right] . \tag{4.51}$$

Note that this is a formal way of writing the action, since $1/(i(v \cdot D) + 2m_Q - i\epsilon)$ is a non-local distribution. This Gaussian integration corresponds to the replacement

$$H_v = \left(\frac{1}{2m_Q + ivD} \right) i\not{D}_\perp h_v \tag{4.52}$$

for the lower component field. The Gaussian integration yields a determinant Δ. In the full theory, one may also perform this Gaussian integration, and the determinant obtained contains all closed loops of heavy quarks. After renormalization of the full theory, their contribution starts at order $1/m^2$ with a term corresponding to the leading contribution to the Uehling potential [33]. In the effective theory, one may take the determinant Δ to be a constant, if the terms of order $1/m_Q^2$ and higher coming from the closed heavy quark loops are included by matching to the full theory. Since we shall discuss only the leading term of the $1/m_Q$ expansion in this section, we may drop the determinant in what follows.

The non-locality of the action functional is connected to the large scale set by the heavy-quark mass, and the non-local terms may be expanded in terms of an infinite series of local operators, which have increasing powers of $1/m_Q$. In the context of a field theory, this corresponds to a short-distance expansion and hence these operators have to be renormalized. The tree-level relations may be read off from the geometric-series expansion of the non-local term in (4.51). In this way, we obtain the expansions of the field and the Lagrangian

$$\begin{aligned} Q(x) &= e^{-im_Q vx}\left[1+\left(\frac{1}{2m+ivD}\right)i\not{D}_\perp\right]h_v \\ &= e^{-im_Q vx}\left[1+\frac{1}{2m_Q}\not{D}_\perp+\left(\frac{1}{2m_Q}\right)^2(-ivD)\not{D}_\perp+\cdots\right]h_v\,, \qquad (4.53) \end{aligned}$$

$$\begin{aligned} \mathcal{L} &= \bar{h}_v(ivD)h_v+\bar{h}_v i\not{D}_\perp\left(\frac{1}{2m+ivD}\right)i\not{D}_\perp h_v \\ &= \bar{h}_v(ivD)h_v+\frac{1}{2m}\bar{h}_v(i\not{D}_\perp)^2 ih_v+\left(\frac{1}{2m}\right)\bar{h}_v(i\not{D}_\perp)(-ivD)(i\not{D}_\perp)h_v \\ &\quad +\cdots\,. \qquad (4.54) \end{aligned}$$

The two expressions (4.54) and (4.53) can be used to express any matrix element involving heavy-quark fields and heavy-quark states as an expansion in $1/m_Q$. As an example, consider a matrix element of a current $\bar{q}\Gamma Q$ mediating a transition between a heavy meson and some arbitrary state $|A\rangle$. Using the expansion of the full QCD field (4.53) and the corresponding expansion of the Lagrangian (4.54), we have, up to order $1/m_Q$,

$$\begin{aligned} \langle A|\bar{q}\Gamma Q|M(v)\rangle &= \langle A|\bar{q}\Gamma h_v|H(v)\rangle+\frac{1}{2m_Q}\langle A|\bar{q}\Gamma P_- i\not{D}h_v|H(v)\rangle \\ &\quad -i\int d^4x\langle A|T\{L_1(x)\bar{q}\Gamma h_v\}|H(v)\rangle+\mathcal{O}(1/m^2)\,, \qquad (4.55) \end{aligned}$$

where L_1 is the $1/m$ corrections to the Lagrangian as given in (4.54). In addition, $|M(v)\rangle$ is the state of the heavy meson in full QCD, including all of its mass dependence, while $|H(v)\rangle$ is the corresponding state in the infinite-mass limit.

Equation (4.55) displays the generic structure of the higher-order corrections as they appear in any HQET calculation. There will be local contributions coming from the expansion of the full QCD field; these may be interpreted as the corrections to the currents. The non-local contributions, i.e. the time-ordered products, are the corrections to the states and thus, in the right-hand side of (4.55), only the states of the infinite-mass limit appear.

The derivation given above is not the only possible way to construct the infinite-mass limit. Another possibility is to use the so called Foldy–Wouthuysen transformation [34], which is used to construct the non-relativistic limit of the Dirac equation. Using this transformation [35] in fact yields a different expansion from the Lagrangian as the one given in (4.54), but the expansion of the fields also turns out to be different from (4.53). However, there can only be a single $1/m$ expansion of a physical matrix element such as (4.55), and it has been shown in [35] that the final result for a physical matrix element is the same in both approaches. This means that terms can be reshuffled from the Lagrangian to the fields without changing the physical content of the theory.

4.4 Heavy-Quark Symmetries

The main impact of the heavy-quark limit is due to two additional symmetries which are not present in full QCD [20, 22]. These symmetries restrict the long-distance contributions in a model-independent way. The first symmetry is the heavy-flavour symmetry. The interaction of the quarks with the gluons is flavour independent; all flavour dependence in QCD is due only to the different quark masses. To leading order in $1/m$, the Lagrangian (4.54) is mass-independent and hence a flavour symmetry relating heavy quarks moving with the same velocity appears.

For the case of two flavours b and c we have, to leading order, the Lagrangian

$$\mathcal{L}_{heavy} = \bar{b}_v (v \cdot D) b_v + \bar{c}_v (v \cdot D) c_v \ , \tag{4.56}$$

where b_v and c_v are the field operator h_v for the b and c quarks, respectively. This Lagrangian is obviously invariant under the $SU(2)_{HF}$ rotations

$$\begin{pmatrix} b_v \\ c_v \end{pmatrix} \to U_v \begin{pmatrix} b_v \\ c_v \end{pmatrix} , \quad U \in SU(2)_{HF}. \tag{4.57}$$

We have put a subscript v on the transformation matrix U, since this symmetry relates only heavy quarks moving with the same velocity.

The second symmetry is the heavy-quark spin symmetry. As is clear from the Lagrangian in the heavy-mass limit, both spin degrees of freedom of the heavy quark couple in the same way to the gluons; we may rewrite the leading-order Lagrangian as

$$\mathcal{L} = \bar{h}_v^{+s} (ivD) h_v^{+s} + \bar{h}_v^{-s} (ivD) h_v^{-s}, \tag{4.58}$$

where $h_v^{\pm s}$ are the projections of the heavy-quark field on a definite spin direction s,

$$h_v^{\pm s} = \frac{1}{2}(1 \pm \gamma_5 \not{s}) h_v \ , \quad s \cdot v = 0 \ , \quad s^2 = -1 \ . \tag{4.59}$$

This Lagrangian has a symmetry under the rotations of the heavy-quark spin and hence all the heavy-hadron states moving with the velocity v fall into spin-symmetry doublets as $m_Q \to \infty$. In Hilbert space, this symmetry is generated by operators $S_v(\epsilon)$ as

$$[h_v, S_v(\epsilon)] = i\not\epsilon\not v\gamma_5 h_v \ , \tag{4.60}$$

where ϵ, with $\epsilon^2 = -1$, is the rotation axis. The simplest spin-symmetry doublet in the mesonic case consists of the pseudoscalar meson $H(v)$ and the corresponding vector meson $H^*(v,\epsilon)$, since a spin rotation yields

$$\exp\left(iS_v(\epsilon)\frac{\pi}{2}\right)|H(v)\rangle = (-i)|H^*(v,\epsilon)\rangle \ , \tag{4.61}$$

where we have chosen an arbitrary phase to be $(-i)$.

The spin symmetry relation between the pseudoscalar and the vector meson can be implemented by using the representation matrices for these states

$$H(v) = \frac{1}{2}\sqrt{m_H}\gamma_5(\not v - 1) \qquad \text{for the pseudoscalar meson,} \tag{4.62}$$

$$H^*(v,\epsilon) = \frac{1}{2}\sqrt{m_H}\not\epsilon(/v - 1) \qquad \text{for the vector meson,} \tag{4.63}$$

where the two indices of the matrices correspond to the indices of the heavy quark and the light anti-quark, respectively.

Using these matrices allows us to exploit the heavy-quark spin symmetry in a very simple way. As an example, we may derive the analogue of the Wigner–Eckart theorem for the heavy-quark spin symmetry. If $\mathcal{H}(v)$ denotes either $H(v)$ or $H^*(v,\epsilon)$ and if $|\mathcal{H}(v)\rangle$ is the corresponding state, we have for any heavy-to-heavy transition current, using the representation matrices on the right-hand side:

$$\langle\mathcal{H}(v')|\bar{h}_{v'}\Gamma h_v|\mathcal{H}(v)\rangle = \xi(v\cdot v') \ \mathrm{Tr}\ \left\{\overline{\mathcal{H}}(v')\Gamma\mathcal{H}(v)\right\} \ , \tag{4.64}$$

where Γ is some arbitrary combination of Dirac matrices. Equation (4.64) is one of the most important results of heavy-quark symmetry in the mesonic sector, since it relates *every* matrix element of heavy-to-heavy currents between two heavy mesons to a single form factor, called Isgur–Wise function ξ. Note that the Isgur–Wise function is, in a group theoretical language, just the reduced matrix element, which is universal for the whole spin–flavour symmetry multiplet. Furthermore, the trace in (4.64) is the Clebsch–Gordan coefficient, which is entirely determined by the current operator and the states of the multiplet.

Furthermore, since the current

$$j_\mu = \bar{h}_v\gamma_\mu h_v \tag{4.65}$$

generates the heavy-flavour symmetry, we have a normalization statement for the Isgur–Wise function

$$\xi(v\cdot v' = 1) = 1 \tag{4.66}$$

since the generators of a symmetry have to have normalized matrix elements.

Treating both the charm and the bottom quark as heavy static quarks, we find that only a single, normalized form factor describes the exclusive decays $B \to D\ell\bar{\nu}_\ell$ and $B \to D^*\ell\bar{\nu}_\ell$, which opens the possibility of a model-independent determination of the CKM matrix element V_{cb}; we shall discuss this when looking at sample applications.

In the heavy-mass limit, the spin-symmetry partners have to be degenerate and their splitting has to scale as $1/m_Q$. From the Lagrangian given above, we can derive the mass relation for the heavy ground-state mesons up to terms of order $1/m_Q$,

$$m_H = m_Q + \bar{\Lambda} + \frac{1}{2m_Q}\left(\lambda_1 + d_H\lambda_2\right) , \tag{4.67}$$

where $d_H = 3$ for the 0^- meson and $d_H = -1$ for the 1^- meson. The parameters $\bar{\Lambda}$, λ_1 and λ_2 correspond to matrix elements involving higher-order terms that appear in the effective-theory Lagrangian,

$$\bar{\Lambda} = \frac{\langle 0|q\, iv\overleftarrow{D}\, \gamma_5 h_v|H(v)\rangle}{\langle 0|q\gamma_5 h_v|H(v)\rangle} , \tag{4.68}$$

$$\lambda_1 = \frac{\langle H(v)|\bar{h}_v(iD)^2 h_v|H(v)\rangle}{2M_H} , \tag{4.69}$$

$$\lambda_2 = \frac{\langle H(v)|\bar{h}_v\sigma_{\mu\nu}iD^\mu iD^\nu h_v|H(v)\rangle}{2M_H} , \tag{4.70}$$

where the normalization of the states has been chosen to be $\langle H(v)|\bar{h}_v h_v|H(v)\rangle = 2M_H = 2(m_Q + \bar{\Lambda})$. These parameters may be interpreted as the binding energy of the heavy meson in the infinite mass limit ($\bar{\Lambda}$), the expectation value of the kinetic energy of the heavy quark (λ_1), and its energy due to the chromomagnetic moment of the heavy quark (λ_2) inside the heavy meson. The latter two parameters play an important role since they parametrize the non-perturbative input needed in the subleading order of the $1/m_Q$ expansion.

The prediction (4.67) from spin symmetry can be checked against data. We have

$$m_{H^*}^2 - m_H^2 \approx 2m_Q(m_{H^*} - m_H) = 4\lambda_2 , \tag{4.71}$$

which yields $\lambda_2 = 0.12\,\mathrm{GeV}$ for the data for both the B and the D meson systems. However, it must be considered an accident that one obtains similar results for the light quark sector also: the mass differences $m_{K^*}^2 - m_K^2$ and $m_\rho - m_\pi$ are consistent with the numbers obtained from the heavy mesons.

Heavy-quark symmetries also allow us to make statements about the behaviour of certain matrix elements once one introduces an explicit breaking of the symmetry, such as the presence of $1/m$ terms. One of the most important results related to the $1/m$ corrections is called Luke's theorem [36]. It is a generalization of the Ademollo–Gatto theorem [37], which states that in the presence of explicit symmetry breaking, the matrix elements of the currents that generate the symmetry are still normalized up to terms which are second-order in the symmetry-breaking interaction.

For the case at hand, the relevant symmetry is the heavy-flavor symmetry. This symmetry is an $SU(2)$ symmetry and is generated by three operators $Q_\pm$ and Q_3, where

$$\begin{aligned}
Q_+ &= \int d^3x\, \bar{b}_v(x) c_v(x) \,, \quad Q_- = \int d^3x\, \bar{c}_v(x) b_v(x) \,, \\
Q_3 &= \int d^3x\, (\bar{b}_v(x) b_v(x) - \bar{c}_v(x) c_v(x)) \,, \\
[Q_+, Q_-] &= Q_3 \,, \qquad [Q_+, Q_3] = -2Q_+ \,, \qquad (Q_+)^\dagger = Q_- \,. \qquad (4.72)
\end{aligned}$$

Let us denote the ground-state flavour symmetry multiplet by $|B\rangle$ and $|D\rangle$. The operators then act in the following way:

$$\begin{aligned}
&Q_3|B\rangle = |B\rangle \,, \qquad Q_3|D\rangle = -|D\rangle \,, \\
&Q_+|D\rangle = |B\rangle \,, \qquad Q_-|B\rangle = |D\rangle \,, \\
&Q_+|B\rangle = Q_-|D\rangle = 0 \,. \qquad (4.73)
\end{aligned}$$

The Hamiltonian of this system has a $1/m_Q$ expansion of the form

$$\begin{aligned}
H &= H_0^{(b)} + H_0^{(c)} + \frac{1}{2m_b} H_1^{(b)} + \frac{1}{2m_c} H_1^{(c)} + \cdots \\
&= H_0^{(b)} + H_0^{(c)} + \frac{1}{2}\left(\frac{1}{2m_b} + \frac{1}{2m_c}\right)(H_1^{(b)} + H_1^{(c)}) \\
&\qquad + \frac{1}{2}\left(\frac{1}{2m_b} - \frac{1}{2m_c}\right)(H_1^{(b)} - H_1^{(c)}) + \cdots \\
&= H_{symm} + H_{break} \,. \qquad (4.74)
\end{aligned}$$

In (4.74), the first line is still symmetric under heavy-flavour $SU(2)$, while the term in the second line does not commute any more with $Q_\pm$, but still commutes with Q_3. In other words, to order $1/m_Q$ we still have common eigenstates of H and Q_3, which we shall denote by $|\tilde{B}\rangle$ and $|\tilde{D}\rangle$. Sandwiching the commutation relation, we obtain

$$\begin{aligned}
1 &= \langle\tilde{B}|Q_3|\tilde{B}\rangle = \langle\tilde{B}|[Q_+, Q_-]|\tilde{B}\rangle \\
&= \sum_n \left[\langle\tilde{B}|Q_+|\tilde{n}\rangle\langle\tilde{n}|Q_-|\tilde{B}\rangle - \langle\tilde{B}|Q_-|\tilde{n}\rangle\langle\tilde{n}|Q_+|\tilde{B}\rangle\right] \\
&= \sum_n \left[|\langle\tilde{B}|Q_+|\tilde{n}\rangle|^2 - |\langle\tilde{B}|Q_-|\tilde{n}\rangle|^2\right] \,, \qquad (4.75)
\end{aligned}$$

where the $|\tilde{n}\rangle$ form a complete set of states of the Hamiltonian $H_{symm} + H_{break}$. The matrix elements may be written as

$$\langle\tilde{B}|Q_\pm|\tilde{n}\rangle = \frac{1}{E_B - E_n}\langle\tilde{B}|[H_{break}, Q_\pm]|\tilde{n}\rangle \,, \qquad (4.76)$$

where E_B and E_n are the energies of the states $|\tilde{B}\rangle$ and $|\tilde{n}\rangle$, respectively. In the case $|\tilde{n}\rangle = |\tilde{D}\rangle$ the matrix element on the left-hand side will be of order unity, since both the numerator and the energy difference in the denominator are of the order of the symmetry breaking. For all other states, the energy difference in the denominator is non-vanishing in the symmetry limit, and hence this difference is of order unity; thus the matrix element for these states will be of the order of the symmetry breaking. From this we conclude that

$$\langle\tilde{B}|Q_+|\tilde{D}\rangle = 1 + \mathcal{O}\left[\left(\frac{1}{2m_b} - \frac{1}{2m_c}\right)^2\right] . \tag{4.77}$$

In particular, the weak transition currents at the non-recoil point $v = v'$ are proportional to these symmetry generators and hence we may conclude that matrix elements that are related in the way shown above to the symmetry generators, can have corrections only of the order $1/m_Q^2$. We shall return to this point when we discuss the applications of the heavy-mass expansion.

Another kind of symmetry is due to the fact that full QCD has Lorentz invariance, while the Lagrangian of HQET does not have this invariance any more. This is obvious, since we have chosen a fixed vector v, which corresponds to a choice of a specific coordinate frame. Only if we were to transform the vector v "by hand" would we obtain invariance again.

On the other hand, when we started from the full QCD Lagrangian, there was no dependence on the vector v, and even when v appeared explicitly in the equations everything was still independent of v, as long as all orders in $1/m_Q$ were taken into account. In other words, the dependence on v emerges at the point where we truncate the $1/m_Q$ expansion.

Hence the $1/m_Q$ expansion has to exhibit an invariance under infinitesimal changes of the vector v, which is called the reparametrization invariance [38, 39]. Reparametrization connects different orders of the $1/m_Q$ expansion and hence there will be relations between coefficients of different orders.

In order to explore the consequences of reparametrization invariance, it is convenient to use the representation (4.54) and (4.53), since we may obtain closed expressions for all orders in the $1/m_Q$ expansion. In fact, by performing an infinitesimal shift of the velocity, combined with the corresponding shift of the covariant derivative and the fields, one can check explicitly that the Lagrangian is invariant under the transformation

$$\begin{aligned} & v \to v + \delta v \,, \qquad v \cdot \delta v = 0 \,, \\ & h_v \to h_v + \frac{\delta \not{v}}{2}\left(1 + P_- \frac{1}{2m_Q + ivD} i\not{D}\right) h_v \,, \\ & iD \to iD - m_Q\, \delta v \,. \end{aligned} \tag{4.78}$$

Reparametrization invariance has to survive renormalization, which means that the relations between coefficients of the Lagrangian derived from (4.78) hold true beyond tree level. One result is the relation between the renormalization of the leading-order Lagrangian and that of the subleading tems, which implies that the kinetic-energy operator $\bar{h}_v (iD)^2 h_v$ is not renormalized to all orders.

4.5 Heavy-Quark Expansion for Inclusive Decays

Inclusive decays of heavy hadrons can also be described using effective-field-theory methods [40, 41, 42, 43, 44, 45].[6] The method is set up in close analogy to deep inelastic scattering and relies on the operator product expansion [8]. The result is an expansion in inverse powers of the heavy-quark mass for inclusive rates and also for spectra. We shall discuss applications of this method in some detail in later chapters; here we shall only outline the theoretical ingredients.

The effective Hamiltonian for a transition in which a heavy flavour (represented by the quark field Q) changes by one unit contains a single Q-quark field, the mass of which sets a large scale. Thus the effective Hamiltonian takes the form

$$\mathcal{H}_{eff} = \bar{Q} R \,, \tag{4.79}$$

where R is other field operators of, for example, light quarks, photons or leptons.

The inclusive decay rate for a heavy hadron H containing the quark Q may be related, via unitarity and the optical theorem, to a forward matrix element by

$$\begin{aligned} \Gamma &\propto \sum_X (2\pi)^4 \delta^4 (P_B - P_X) |\langle X | \mathcal{H}_{eff} | H(v) \rangle|^2 \\ &= \int d^4x \, \langle H(v) | \mathcal{H}_{eff}(x) \mathcal{H}^\dagger_{eff}(0) | H(v) \rangle \\ &= 2 \ \mathrm{Im} \int d^4x \, \langle H(v) | T\{ \mathcal{H}_{eff}(x) \mathcal{H}^\dagger_{eff}(0) \} | H(v) \rangle \,, \end{aligned} \tag{4.80}$$

where $|X\rangle$ is the final state, which is summed over to obtain the inclusive rate.

In order to exploit the fact that m_Q is a scale large compared with Λ_{QCD}, we perform the same field redefinition as we did when deriving the Lagrangian for HQET (see (4.46)),

$$Q_v = e^{-im_Q (v \cdot x)} Q \,. \tag{4.81}$$

This leads to

$$\Gamma \propto 2 \ \mathrm{Im} \int d^4x \, e^{-im_Q vx} \, \langle H(v) | T\{ \widetilde{\mathcal{H}}_{eff}(x) \widetilde{\mathcal{H}}^\dagger_{eff}(0) \} | H(v) \rangle \,, \tag{4.82}$$

---

[6] Early work on inclusive decays and lifetimes, which actually pre-dates the heavy-quark expansion, can be found in [46, 47, 48].

where

$$\widetilde{\mathcal{H}}_{eff} = \bar{Q}_v R \, . \tag{4.83}$$

This relation exhibits the similarity between cross-section calculation for deep inelastic scattering and this approach to total rates. In deep inelastic scattering, there appears a large scale, which is the momentum transfer q to the leptons, while here the mass of the heavy quark appears as a large scale. However, in deep inelastic scattering the momentum transfer is in the deep Euclidean region $-q^2 \gg \Lambda_{QCD}$, while in the case of the decay of a heavy hadron this vector is in the Minkowskian region $(m_Q v)^2 \gg \Lambda_{QCD}$. A strict proof of the operator product expansion exists only in the deep Euclidean region, and the analytic continuation to the Minkowskian region could introduce problems, which have been discussed recently in the context of duality violations. We shall not discuss this question here any further; rather, we refer the interested reader to recent reviews [49, 50, 51].

After the phase redefinition, the remaining matrix element does not involve large momenta of the order of the heavy-quark mass any more, and hence a short-distance expansion becomes useful if the mass m_Q is large compared with the scale $\bar{\Lambda}$ determining the matrix element. The next step is thus to perform an operator product expansion, which has the general form

$$\int d^4x \, e^{i m_Q v x} \, T\{\widetilde{\mathcal{H}}_{eff}(x)\widetilde{\mathcal{H}}^\dagger_{eff}(0)\} = \sum_{n=0}^{\infty} \left(\frac{1}{2m_Q}\right)^n \hat{C}_{n+3}(\mu)\mathcal{O}_{n+3}(\mu) \, , \tag{4.84}$$

where the $\mathcal{O}_n$ are operators of dimension n, with their matrix elements renormalized at scale μ, and $\hat{C}_n$ are the corresponding Wilson coefficients.

In order to compute the total rate, we have to take a forward matrix element with the decaying heavy hadron, i.e.

$$\Gamma \propto 2 \, \mathrm{Im} \sum_{n=0}^{\infty} \left(\frac{1}{2m_Q}\right)^n \hat{C}_{n+3}(\mu)\langle H(v)|\mathcal{O}_{n+3}(\mu)|H(v)\rangle, \tag{4.85}$$

which shows that this expansion still does not yield the full expansion in inverse powers of the heavy mass, since the state $|H(v)\rangle$ is that of full QCD and thus still has a dependence on the heavy mass. In order to obtain the complete expansion in inverse powers of the heavy mass, we have to use the methods of HQET as described in the last section and expand every matrix element in $1/m_Q$. However, it has been argued that it is in fact advantageous to omit the HQET expansion and to treat the matrix elements as phenomenological parameters [52].

The lowest-order terms of the operator product expansion are the dimension-three operators. Owing to Lorentz invariance and parity there are only two combinations which can appear, namely $\bar{Q}_v \not{v} Q_v$ and $\bar{Q}_v Q_v$. Note that the operators Q_v differ from the full QCD operators only by a phase redefinition, and hence $\bar{Q}_v \not{v} Q_v = \bar{Q} \not{v} Q$ and $\bar{Q}_v Q_v = \bar{Q}Q$. The first combination is proportional to the Q-number current $\bar{Q}\gamma_\mu Q$, which is normalized even in full QCD, while the second combination differs from the first one only by terms of order $1/m_Q^2$:

$$\bar{Q}_v Q_v = v^\mu \bar{Q}_v \gamma_\mu Q_v + \frac{1}{2m_Q^2} \bar{Q}_v \left[(iD)^2 - (ivD)^2 + \frac{i}{2}\sigma_{\mu\nu}G^{\mu\nu} \right] Q_v + \mathcal{O}(1/m_Q^3) , \tag{4.86}$$

where $G_{\mu\nu}$ is the gluon field strength.

Thus the matrix elements of the dimension-three contribution are known; in the standard normalization of the states this implies

$$\langle H(v)|\mathcal{O}_3|H(v)\rangle = \langle H(v)|\bar{Q}_v \not{v} Q_v|B(v)\rangle = 2m_H , \tag{4.87}$$

where m_H is the mass of the heavy hadron. To lowest order in the heavy-mass expansion we may furthermore replace $m_B = m_Q$ and hence we may evaluate the leading term in the $1/m_q$ expansion without any hadronic uncertainty. The evaluation of the corresponding Wilson coefficient yields the result that this coefficient is the free quark decay rate. In this way the naive ansatz, namely that of using the decay rate of a free heavy quark (i.e. the parton model) as an approximation to the total decay rate of a heavy hadron, turns out to be the leading term of a $1/m_Q$ expansion.

A dimension-four operator contains an additional covariant derivative, and thus we have matrix elements of the type

$$\langle H(v)|\mathcal{O}_4|H(v)\rangle \propto \langle H(v)|\bar{Q}_v \Gamma D_\mu Q_v|H(v)\rangle = \mathcal{A}_\Gamma v_\mu . \tag{4.88}$$

Since the equations of motion apply to this tree-level matrix element, we find that the constant $\mathcal{A}_\Gamma$ has to vanish, and thus there are no dimension-four contributions. This statement is completely equivalent to Luke's theorem [36], since we are considering a forward matrix element, i.e. a matrix element at zero recoil [53].

The first non-trivial non-perturbative contributions come from dimension-five operators and are thus of order $1/m_Q^2$. For mesonic decays there are only the two parameters λ_1 and λ_2 given in (4.69) and (4.70), which correspond to matrix elements of the subleading terms of the Lagrangian. They parametrize the non-perturbative input at order $1/m_Q^2$. For Λ_Q-type baryons the parameter λ_2 vanishes owing to heavy-quark spin symmetry, while the kinetic-energy parameter λ_1 is nonzero. In the framework of the $1/m_Q$ expansion, this leads to a difference in lifetimes between mesons and baryons, which we shall discuss in Sect. 5.4.

We can discuss differential rates along the same lines. However, here the operator product expansion as outlined above is applied not to the full effective Hamiltonian, but only to the hadronic currents. As an example, we discuss semileptonic decays of a heavy hadron. In this case the differential rate is written as a product of the hadronic and leptonic tensors

$$d\Gamma = \frac{G_F^2}{4m_B}|V_{Qq}|^2 W_{\mu\nu}\Lambda^{\mu\nu}d(PS) \, , \tag{4.89}$$

where $d(PS)$ is the phase-space differential. The phase redefinition (4.81) of the heavy-quark fields now yields the momentum-transfer variable

$$Q \equiv m_Q v - q \, , \tag{4.90}$$

where q is the momentum transferred to the leptons. The variables Q^2 and $(v \cdot Q)^2$ (v is the velocity of the decaying heavy hadron) have to be large compared with Λ_{QCD}^2 in order to justify the short-distance expansion.

The structure of the expansion of the spectrum is identical to that for the total rate. The contribution of the dimension-three operators yields the free-quark decay spectrum, there are no contributions from dimension-four operators, and the $1/m_b^2$ corrections are parametrized in terms of λ_1 and λ_2. Calculating the spectrum for $B \to X_c \ell \nu$ yields [44, 45]

$$\begin{aligned}\frac{d\Gamma}{dy} &= \frac{G_F^2 \mid V_{cb}|^2 m_Q^5}{192\pi^3}\Theta(1-y-\rho)y^2 \Bigg[\{3(1-\rho)(1-R^2) - 2y(1-R^3)\} \\ &+\frac{\lambda_1}{[m_Q(1-y)]^2}(3R^2-4R^3) - \frac{\lambda_1}{m_Q^2(1-y)}(R^2-2R^3) \\ &-\frac{3\lambda_2}{m_Q^2(1-y)}(2R+3R^2-5R^3) + \frac{\lambda_1}{3m_Q^2}[5y-2(3-\rho)R^2+4R^3] \\ &+\frac{\lambda_2}{m_Q^2}[(6+5y)-12R-(9-5\rho)R^2+10R^3]\Bigg] + \mathcal{O}\left[(\Lambda/[m_Q(1-y)])^3\right]\end{aligned} \tag{4.91}$$

where we have defined

$$\rho = \left(\frac{m_f}{m_Q}\right)^2 \, , \quad R = \frac{\rho}{1-y} \, , \tag{4.92}$$

assuming that the final-state quark has the mass m_f, and

$$y = 2E_\ell/m_b \tag{4.93}$$

is the rescaled energy of the charged lepton.

This expression is somewhat complicated, but it simplifies for the case $m_f = 0$. One finds

$$\begin{aligned}\frac{d\Gamma}{dy} &= \frac{G_F^2 \mid V_{ub}|^2 m_Q^5}{192\pi^3}\left[\left(2y^2(3-2y) + \frac{10y^2}{3}\frac{\lambda_1}{m_Q^2} + 2y(6+5y)\frac{\lambda_2}{m_Q^2}\right)\Theta(1-y)\right. \\ &\left. -\frac{\lambda_1+33\lambda_2}{3m_Q^2}\delta(1-y) - \frac{\lambda_1}{3m_Q^2}\delta'(1-y)\right] \, .\end{aligned} \tag{4.94}$$

It is obvious from (4.91) and (4.94) that the spectra behave pathologically in the endpoint region. This is to be expected, since the expansion parameter for semileptonic decay is not really $1/m_Q$, but rather is $1/[m_Q(1-y)]$, indicating that the expansion breaks down in the region $y \approx 1$.

This problem is even more pronounced in the case of the inclusive process $B \to X_s\gamma$. The leading term is given by the partonic rate of $b \to s\gamma$, which is a two-particle decay at tree level. Thus the energy spectrum of the photon is a δ-function

$$\frac{d\Gamma}{dx} = \frac{G_F^2 \alpha m_b^5}{32\pi^4} |V_{ts}V_{tb}^*|^2 |C_7|^2 \delta(1-x) \,, \tag{4.95}$$

where $x = 2E_\gamma/m_b$ and C_7 is the Wilson coefficient of the effective Hamiltonian, which is described in some detail in the next section.

Including subleading terms in the $1/m_b$ expansion (but still working at tree level in the α_s expansion) does not change the fact that the final state is a two-particle state; hence the corrections are still given by local distributions and read

$$\frac{d\Gamma}{dx} = \frac{G_F^2 \alpha m_b^5}{32\pi^4} |V_{ts}V_{tb}^*|^2 |C_7|^2 \times \left(\delta(1-x) - \frac{\lambda_1 + 3\lambda_2}{2m_b^2}\delta'(1-x) + \frac{\lambda_1}{6m_b^2}\delta''(1-x) + \cdots\right) \,, \tag{4.96}$$

A non-trivial spectrum is obtained only after including the emission of a real gluon, in which case the hadronic invariant mass of the final state is non-zero and the photon spectrum extends over all the kinematically allowed region.

It has been shown that these singular terms can be resummed using a slightly different expansion, the called the twist expansion, which we shall discuss in the next section.

4.6 Twist Expansion for Heavy-Hadron Decays

As we have seen in the last section, the endpoint regions of inclusive semileptonic decays (i.e. the regions in which $E_\ell \sim E_{\max}$) and the endpoint region in the inclusive process $B \to X_s\gamma$ (which is $E_\gamma \sim E_{\max}$) exhibit a singular behaviour. This can be traced back to the fact that the expansion parameter for spectra of inclusive decays is not $1/m_b$; rather, it is $1/[m_Q(1-y)] = 1/(m_b - 2E_\ell)$ for the semileptonic decay and $1/[m_Q(1-x)] = 1/(m_b - 2E_\gamma)$ for the decay $B \to X_s\gamma$ (see (4.91)). Close to the endpoint, these expansion parameters become large and the expansion breaks down.

In order to cure this problem, it is instructive to study $B \to X_s\gamma$. The general structure of the rate for this decay at tree level (but including $1/m_b$ corrections) is

$$\frac{d\Gamma}{dx} = \frac{G_F^2 \alpha m_b^5}{32\pi^4} |V_{ts} V_{tb}^*|^2 |C_7|^2$$
$$\times \left[\sum_i a_i \left(\frac{1}{m_b} \right)^i \delta^{(i)}(1-x) + \mathcal{O}\{(1/m_b)^{i+1} \delta^{(i)}(1-x)\} \right] , \quad (4.97)$$

where $\delta^{(n)}$ denotes the nth derivative of the δ-function.

The kinematics in the endpoint regions requires a different expansion [54, 55, 56, 57], which is analogous to the expansion performed in deep inelastic scattering. From the kinematics we have $p_B = M_B v = p_X + q$, where p_X is the hadronic momentum of the final state and q is the momentum of the photon. In the case of semileptonic decay, p_X would be the sum of the momenta of the final-state hadrons and the neutrino, while q would be the momentum of the charged lepton. The endpoint region is defined by the region where the hadronic invariant mass is small, of order $\sqrt{\Lambda_{QCD} m_b}$ while the hadronic energy is still large of order m_b:

$$(p_B - q)^2 \sim \mathcal{O}(\Lambda_{QCD} m_b) , \qquad M_B - v \cdot q \sim \mathcal{O}(m_b) . \quad (4.98)$$

It is convenient to introduce light-cone vectors n_+ and n_- in the form

$$q = \frac{1}{2}(n_+ \cdot q) n_- , \qquad v = \frac{1}{2}(n_+ + n_-) , \quad (4.99)$$

such that we may decompose every vector P as

$$P = \frac{1}{2}(n_+ \cdot P) n_- + \frac{1}{2}(n_- \cdot P) n_+ + P_\perp . \quad (4.100)$$

Using these definitions we may now discuss the expansion of the correlator

$$R = \int d^4x \exp(-ix[m_b v - q]) \langle B(p_B) | \bar{b}_v(0) \Gamma q(0) \bar{q}(x) \Gamma^\dagger b_v(x) | B(p_B) \rangle , \quad (4.101)$$

where we have already performed the phase redefinition (4.46) of the heavy-quark field. The usual $1/m_b$ expansion is recovered by performing a short-distance expansion of the matrix element, i.e. assuming $x_\mu \to 0$. However, in the case at hand we have in the exponent

$$x[m_b v - q] = \frac{m}{2}(x n_+) + \frac{1}{2}(m_b - nq)(x n_-) . \quad (4.102)$$

Owing to the kinematics in the endpoint region, we have $m_b - n_+ q = m_b - 2(vq) \sim \Lambda_{\text{QCD}}$ such that the first term dominates. Non-vanishing contributions are picked up only in a region where $(xn) = \mathcal{O}(1/m_b)$, which is the region close to the light cone.

Close to the light cone we can perform perturbative calculations and hence we contract the light-quark propagator using the leading-order perturbative result

$$R = \int d^4x \int \frac{d^4Q}{(2\pi)^4} \Theta(Q_0)(2\pi)\delta(Q^2) \exp(-ix[m_b v - q - Q])$$
$$\times \langle B(v)|\bar{b}_v(0)\Gamma \not{Q} \Gamma^\dagger b_v(x)|B(v)\rangle \ , \tag{4.103}$$

where now $|B(v)\rangle$ is the static B meson state. Performing a (gauge-covariant) Taylor expansion of the remaining x dependence of the matrix element, we obtain

$$\begin{aligned} R &= \int d^4x \int \frac{d^4Q}{(2\pi)^4} \Theta(Q_0)(2\pi)\delta(Q^2) \exp(-ix[m_b v - q - Q]) \\ &\quad \times \sum_n \frac{1}{n!} \langle B(v)|\bar{b}_v(0)\Gamma \not{Q} \Gamma^\dagger (-ix \cdot iD)^n b_v(0)|B(v)\rangle \\ &= \int d^4x \int \frac{d^4Q}{(2\pi)^4} \Theta(Q_0)(2\pi)\delta(Q^2) \\ &\quad \times \langle B(v)|\bar{b}_v(0)\Gamma \not{Q} \Gamma^\dagger \exp(-ix[m_b v - q - Q + iD]) b_v(0)|B(v)\rangle \ . \end{aligned} \tag{4.104}$$

When we use spin symmetry and the usual representation matrices of the 0^- B meson states, the matrix element which appears in (4.104) becomes

$$\begin{aligned} &\langle B(v)|\bar{b}_v(0)\Gamma \not{Q} \Gamma^\dagger (iD_{\mu_1})(iD_{\mu_2}) \cdots (iD_{\mu_n}) b_v(0)|B(v)\rangle \\ &\quad = \frac{M_B}{2} \mathrm{Tr}\{\gamma_5(\not{v}+1)\Gamma \not{Q} \Gamma^\dagger (\not{v}+1)\gamma_5\} \\ &\qquad \times [a_1^{(n)} v_{\mu_1} v_{\mu_2} \cdots v_{\mu_n} + a_2^{(n)} g_{\mu_1\mu_2} v_{\mu_3} \cdots v_{\mu_n} + \cdots] \ , \end{aligned} \tag{4.105}$$

where the ellipses denote terms with one or more $g_{\mu\nu}$'s and also antisymmetric terms, and the $a_i^{(n)}$ are non-perturbative parameters. When this result is contracted with $x^{\mu_1} x^{\mu_2} \cdots x^{\mu_n}$, all antisymmetric contributions vanish; furthermore, since the relevant kinematics restricts the x_μ to be on the light cone, also all the $g_{\mu\nu}$ terms are suppressed relative to the first term, which has only v_μ's. Hence

$$\begin{aligned} &\langle B(v)|\bar{b}_v(0)\Gamma \not{Q} \Gamma^\dagger (-ix \cdot iD)^n b_v(0)|B(v)\rangle \\ &\quad = M_B \mathrm{Tr}\{(\not{v}+1)\Gamma \not{Q} \Gamma^\dagger\} a_1^{(n)} (v \cdot x)^n \ , \end{aligned} \tag{4.106}$$

In this way we have made explicit the fact that the kinematics we are studying here forces us to resum the series in $1/m_b$. Defining the twist t of an operator $\mathcal{O}$ in the usual way $t = \dim[\mathcal{O}] - \ell$, where ℓ is the spin of the operator, we find that the resummation corresponds to the contributions of leading twist, $t = 3$. Since x_μ is light-like, it projects out only the light-cone component $n_+ D = D_+$ of the covariant derivative in (4.106). Hence we may write $a_1^{(n)}$ as

$$2M_B a_1^{(n)} = \langle B(v)|\bar{b}_v (iD_+)^n b_v|B(v)\rangle \ , \tag{4.107}$$

and one obtains as a final result

$$R = \int d^4x \int \frac{d^4Q}{(2\pi)^4} \Theta(Q_0)(2\pi)\delta(Q^2)\frac{1}{2}\mathrm{Tr}\{(\not{v}+1)\Gamma \not{Q} \Gamma^\dagger\}$$
$$\langle B(v)|\bar{b}_v \exp(-ix[(m_b + iD_+)v - q - Q])b_v|B(v)\rangle \,. \tag{4.108}$$

Introducing the shape function (or light-cone distribution function) in the form [54, 56, 57]

$$2M_B f(k_+) = \langle B(v)|\bar{b}_v \delta(k_+ - iD_+)b_v|B(v)\rangle \,, \tag{4.109}$$

we can write the result as

$$R = \int dk_+ f(k_+) \int \frac{d^4Q}{(2\pi)^4} \Theta(Q_0)(2\pi)\delta(Q^2)$$
$$\times M_B \mathrm{Tr}\{(\not{v}+1)\Gamma \not{Q} \Gamma^\dagger\}(2\pi)^4 \delta^4([m_b + k_+]v - Q - q) \,. \tag{4.110}$$

Using the shape function corresponds to a resummation of the contributions of leading twist. This is analogous to what is done in deep inelastic scattering, where the parton distribution functions correspond to the shape function f. The only difference, which is already true for the usual $1/m_b$ expansion, is that the the operator product expansion to set up the heavy-quark mass expansion is performed in the Minkowskian region. While in deep inelastic scattering the momentum transfer to the proton is at $q^2 \to -\infty$, one has to continue analytically to $q^2 = m_b^2 > 0$ in heavy-quark physics.

Subleading twist contributions were originally discussed in [58] and have been applied to semileptonic decays in [59, 60, 61]. At leading twist, all contributions to R can be rewritten as a convolution of two non-local operators with a Wilson-coefficient function. The operators are given by

$$O_0(\omega) = \bar{h}_v \delta(\omega + iD_+)h_v \tag{4.111}$$
$$P_0^\alpha(\omega) = \bar{h}_v(0)\gamma^\alpha\gamma_5 \delta(\omega + iD_+)h_v \,, \tag{4.112}$$

The forward matrix element of the operator O_0 is the shape function (see (4.109)), while the forward matrix element of the operator P_0^α vanishes for a B meson, but is non-vanishing for a Λ_b baryon, for example.

Expanding $f(\omega)$ in powers of iD_+ gives the series of increasingly singular terms

$$f(\omega) = \delta(\omega) - \frac{\lambda_1}{6m_b^2}\delta''(\omega) - \frac{\rho_1}{18m_b^3}\delta'''(\omega) + \cdots \,, \tag{4.113}$$

where

$$\frac{1}{2m_B}\langle B|\bar{h}_v(iD_\alpha)(iD_\beta)h_v|B\rangle \equiv \frac{1}{3}(g_{\alpha\beta} - v_\alpha v_\beta)\lambda_1 \,, \tag{4.114}$$
$$\frac{1}{2m_B}\langle B|(iD_\alpha)(iD_\mu)(iD_\beta)|B\rangle \equiv \frac{1}{3}(g_{\alpha\beta} - v_\alpha v_\beta)v_\mu\rho_1 \,, \tag{4.115}$$

corresponding to the singular terms in the spectrum.

The convolution takes the form

$$\hat{R} = -\frac{1}{2m_b}\int_{-\infty}^{\infty} d\omega \left[C_0(\omega)O_0(\omega) + C_{5,0}^{\alpha}(\omega)P_{0,\alpha}(\omega) + \mathcal{O}\left(\frac{\Lambda_{\rm QCD}}{m_b}\right)\right] , \quad (4.116)$$

where the R discussed above is the forward matrix element of the operator $\hat{R}$.

The subleading terms can be treated on a similar footing; they are given by forward matrix elements of subleading operators convoluted with Wilson-coeffcient functions. The parity-even operators are

$$\begin{aligned} O_1^{\mu}(\omega) &= \bar{h}_v \left\{iD^{\mu}, \delta(iD_+ + \omega)\right\} h_v \ , \\ O_2^{\mu}(\omega) &= i\bar{h}_v \left[iD^{\mu}, \delta(iD_+ + \omega)\right] h_v \ , \\ O_3^{\mu\nu}(\omega_1, \omega_2) &= \bar{h}_v \delta(iD_+ + \omega_2) \left\{iD_{\perp}^{\mu}, iD_{\perp}^{\nu}\right\} \delta(iD_+ + \omega_1) h_v \ , \\ O_4^{\mu\nu}(\omega_1, \omega_2) &= g\bar{h}_v \delta(iD_+ + \omega_2) G_{\perp}^{\mu\nu} \delta(iD_+ + \omega_1) h_v \ . \end{aligned} \quad (4.117)$$

The parity-odd operators are

$$\begin{aligned} P_{1,\alpha}^{\mu}(\omega) &= \bar{h}_v \left\{iD^{\mu}, \delta(iD_+ + \omega)\right\} \gamma_{\alpha}\gamma_5 h_v \ , \\ P_{2,\alpha}^{\mu}(\omega) &= i\bar{h}_v \left[iD^{\mu}, \delta(iD_+ + \omega)\right] \gamma_{\alpha}\gamma_5 h_v \ , \\ P_{3,\alpha}^{\mu\nu}(\omega_1, \omega_2) &= \bar{h}_v \delta(iD_+ + \omega_2) \left\{iD_{\perp}^{\mu}, iD_{\perp}^{\nu}\right\} \delta(iD_+ + \omega_1) \gamma_{\alpha}\gamma_5 h_v \ , \\ P_{4,\alpha}^{\mu\nu}(\omega_1, \omega_2) &= g\bar{h}_v \delta(iD_+ + \omega_2) G_{\perp}^{\mu\nu} \delta(iD_+ + \omega_1) \gamma_{\alpha}\gamma_5 h_v \ . \end{aligned}$$

Finally, at subleading order there are also contributions from the time-ordered products of $O_0(\omega)$ with the subleading terms in the HQET Lagrangian,

$$\mathcal{O}_{1/m}(y) = \bar{h}_v(y)(iD)^2 h_v(y) + \frac{g}{2}\bar{h}_v(y)\sigma_{\mu\nu}G^{\mu\nu}h_v(y) \, . \quad (4.118)$$

This yields another two operators

$$\begin{aligned} O_T(\omega) &= i\int d^4y \, \frac{1}{2\pi}\int dt \, e^{-i\omega t} T\left(\bar{h}_v(0)h_v(t)\mathcal{O}_{1/m}(y)\right) \\ P_{T,\alpha}(\omega) &= i\int d^4y \, \frac{1}{2\pi}\int dt \, e^{-i\omega t} T\left(\bar{h}_v(0)\gamma_{\alpha}\gamma_5 h_v(t)\mathcal{O}_{1/m}(y)\right) \, . \end{aligned} \quad (4.119)$$

At subleading order, the nonlocal operator product expansion in (4.116) is

$$\begin{aligned} -2m_b\hat{R} = &\int d\omega \, \left(C_0(v,q,\omega)O_0(\omega) + C_{5,0}^{\alpha}(v,q,\omega)P_{0,\alpha}(\omega)\right) \\ &+ \frac{1}{2m_b}\sum_{i=1,2}\int d\omega \, \left(C_i^{\mu}(v,q,\omega)O_{i,\mu}(\omega) + C_{5,i}^{\alpha,\mu}(v,q,\omega)P_{i,\alpha,\mu}(\omega)\right) \end{aligned}$$

$$
\begin{aligned}
&+\frac{1}{2m_b}\sum_{i=3,4}\int d\omega_1 d\omega_2\,\left(C_i^{\mu\nu}(v,q,\omega_1,\omega_2)O_{i,\mu\nu}(\omega_1,\omega_2)\right.\\
&\qquad\qquad\left.+C_{5,i}^{\alpha,\mu\nu}(v,q,\omega_1,\omega_2)P_{i,\alpha,\mu\nu}(\omega_1,\omega_2)\right)\\
&+\frac{1}{2m_b}\int d\omega\,\left(C_T(v,q,\omega)O_T(\omega)+C_{5,T}^{\alpha}(v,q,\omega)P_{T,\alpha}(\omega)\right)\\
&\quad+O\left(\frac{\Lambda_{\rm QCD}^2}{m_b^2}\right)\,. \qquad (4.120)
\end{aligned}
$$

The matching at subleading order onto the operators (4.117) and (4.118) is performed by computing the zero, one and two gluon matrix elements in full QCD and comparing the result with the operators in (4.117) and (4.118). Note that one also has to include the terms from the expansion of the b quark field,

$$
b=\left(1+\frac{i\not{D}}{2m_b}+\dots\right)h_v. \qquad (4.121)
$$

Subleading functions are defined by taking forward matrix elements of these operators. Writing the most general ansatz consistent with the symmetries and the equation of motion $(iv\cdot D)\,h=0$, we find that only the following matrix elements are non-vanishing:

$$
\begin{aligned}
\langle B(v)|O_1^{\mu}(\omega)|B(v)\rangle &= 2m_B g_1(\omega)(v^{\mu}-n_+^{\mu})\,,\\
\langle B(v)|O_3^{\mu\nu}(\omega_1,\omega_2)|B(v)\rangle &= 2m_B g_2(\omega_1,\omega_2)g_{\perp}^{\mu\nu}\,,\\
\langle B(v)|P_{2,\alpha}^{\mu}(\omega)|B(v)\rangle &= 2m_B h_1(\omega)\varepsilon_{\perp,\alpha}^{\mu}\,, \qquad (4.122)\\
\langle B(v)|P_{4,\alpha}^{\mu\nu}(\omega_1,\omega_2)|B(v)\rangle &= 2m_B h_2(\omega_1,\omega_2)\varepsilon_{\rho\sigma\alpha\beta}\,g_{\perp}^{\mu\rho}g_{\perp}^{\nu\sigma}v^{\beta}\,,\\
\langle B(v)|O_T(\omega)|B(v)\rangle &= 2m_B t(\omega)\,, \qquad (4.123)
\end{aligned}
$$

where we have defined

$$
\varepsilon_{\perp}^{\mu\nu}=\varepsilon^{\mu\nu\alpha\beta}v_{\alpha}n_{+\,\beta}\,, \qquad (4.124)
$$

and $\varepsilon^{0123}=1$. Owing to the equations of motion we can eliminate one of these functions, since it is given in terms of the leading-order function:

$$
\begin{aligned}
2m_B g_1(\omega) &= n_{\mu}\langle B(v)|O_1^{\mu}(\omega)|B(v)\rangle=\langle B(v)|\bar{h}_v\left\{iD_+,\delta(iD_++\omega)\right\}h_v|B(v)\rangle\\
&=-2(m_b\,\omega)\langle B(v)|\bar{h}_v\delta(iD_++\omega)h_v|B(v)\rangle=-4m_B(m_b\,\omega)\,f(\omega)\,. \qquad (4.125)
\end{aligned}
$$

For the other subleading functions, we only have information about moments. The moment expansions of the leading and subleading functions read

$$
\begin{aligned}
f(\omega) &= \delta(\omega)-\frac{\lambda_1}{6m_b^2}\delta''(\omega)-\frac{\rho_1}{18m_b^3}\delta'''(\omega)+\cdots\,,\\
\omega f(\omega) &= \frac{\lambda_1}{3m_b^2}\delta'(\omega)+\frac{\rho_1}{6m_b^3}\delta''(\omega)+\cdots\,,\\
h_1(\omega) &= \frac{\lambda_2}{m_b}\delta'(\omega)+\frac{\rho_2}{2m_b^2}\delta''(\omega)+\cdots\,,
\end{aligned}
$$

$$g_2(\omega_1,\omega_2) = \frac{2\lambda_1}{3}\delta(\omega_1)\delta(\omega_2) + \cdots ,$$
$$h_2(\omega_1,\omega_2) = \lambda_2\delta(\omega_1)\delta(\omega_2) + \cdots ,$$
$$t(\omega) = -\frac{\lambda_1 + 3\lambda_2}{m_b}\delta'(\omega) + \frac{\tau}{2m_b^2}\delta''(\omega) + \cdots .$$

We shall make use of these relations and definitions when we consider examples in Sect. 5.4.

It turns out that the two subleading functions g_2 and h_2 appear in the applications to be discussed later only in specific combinations, which are

$$G_2(\omega) = \int d\omega_1\, d\omega_2 \frac{\delta(\omega-\omega_1) - \delta(\omega-\omega_2)}{\omega_1 - \omega_2} g_2(\omega_1,\omega_2) , \qquad (4.126)$$

$$H_2(\omega) = \int d\omega_1\, d\omega_2 \frac{\delta(\omega-\omega_1) - \delta(\omega-\omega_2)}{\omega_1 - \omega_2} h_2(\omega_1,\omega_2) . \qquad (4.127)$$

Furthermore, reparametrization invariance (see the end of Sect. 4.4) can also be used for the light-cone distributions functions [62, 63]. In this case, reparametrization invariance requires that the subleading functions G_2 given in (4.126) and the contribution containing the time-ordered product involving the kinetic-energy term (4.123) appear only in the combination

$$F(\omega) = f(\omega) + \frac{1}{2m_b} G_2(\omega) + t(\omega) , \qquad (4.128)$$

which reduces the number of unknown functions appearing at subleading order to only three.

The light-cone distribution functions can also be derived using the language of effective field theory. In particular, when one wants to consider radiative corrections, an effective-field-theory picture simplifies matters considerably. However, the effective theory that has to be formulated has a few peculiar features, which will be discussed in the next section.

4.7 Soft-Collinear Effective Field Theory

Another effective field theory derived from QCD is soft-collinear effective field theory [64, 65, 66, 67], which is similar to heavy-quark effective theory (HQET). In HQET all the light degrees of freedom have to have momenta of the order Λ_{QCD}, i.e. the momentum p of the heavy quark inside a heavy meson moving with velocity $v = p_{\text{Meson}}/M_{\text{Meson}}$ is decomposed as $p = m_{\text{quark}}v + k$, and all components of the residual momentum k are assumed to be of order Λ_{QCD}.

However, in a decay of a heavy quark into a light quark one may have a kinematical situation in which the light degrees of freedom carry a large energy in the rest frame of the heavy quark, i.e. $vp \sim m_b$, where p is the momentum of the light quark. As an example, one may consider the inclusive decay $B \to X_s\gamma$ in the corner of phase space where the energy E_γ of the outgoing photon is close to its maximal value[7] of $E_{\max} = M_B/2$. In this case the hadronic final state corresponds to a collimated "jet" of hadrons with a small invariant mass but with a large energy in the rest frame of the decaying B meson.

Soft-collinear effective theory (SCET) is designed to describe such a situation. It turned out that an early attempt (called "large-energy effective field theory (LEET) [68]) failed to describe certain degrees of freedom correctly and had to be supplemented. In order to give some idea about SCET, we shall consider again the example of $B \to X_s\gamma$. In this case we have

$$M_B v = m_b v + k = q + p \,, \tag{4.129}$$

where q is the momentum of the photon and p that of the hadronic final state.

As before, we introduce light-cone vectors n_+ and n_- by[8]

$$v = \frac{1}{2}(n_+ + n_-)\,, \quad q = \frac{1}{2}(n_+q)n_-\,, \quad n_\pm^2 = 0\,, \quad (n_-n_+) = 2 \tag{4.130}$$

in terms of which we can decompose the metric as

$$g^{\mu\nu} = \frac{1}{2}(n_+^\mu n_-^\nu + n_-^\mu n_+^\nu) + g_\perp^{\mu\nu}\,. \tag{4.131}$$

Using this, we can write

$$\begin{aligned} m_b v + k - q &= \frac{1}{2}\left([m_b + (n_-k)]n_+ + [m_b + (n_+k) - (n_+q)]n_-\right) + k_\perp \\ &= \frac{1}{2}\left((n_-p)n_+ + (n_+p)n_-\right) + p_\perp\,. \end{aligned} \tag{4.132}$$

In the endpoint region, where the photon energy is close to its maximal value, we have $(n_+q) \sim m_b$ such that $[m_b - (n_+q)] \sim \Lambda_{\rm QCD}$. This means that in this region, the momentum of the final-state hadrons is an almost light-like vector along the n_+ direction:

$$(n_-p) = [m_b + (n_-k)] \sim \mathcal{O}(m_b)\,, \tag{4.133}$$

$$(n_+p) = [m_b - (n_+q) + (n_+k)] \sim \mathcal{O}(\Lambda_{\rm QCD})\,, \tag{4.134}$$

$$p_\perp = k_\perp \sim \mathcal{O}(\Lambda_{\rm QCD})\,. \tag{4.135}$$

However, in order to define a consistent power counting, it is convenient to introduce a dimensionless parameter λ such that the final-state invariant mass is

[7] We ignore the mass of the final state for the moment.

[8] An explicit realization in the frame where $v = (1,0,0,0)$ would be $n_+ = (1,0,0,1)$ and $n_+ = (1,0,0,-1)$.

$$p^2 = (n_+p)(n_-p) + p_\perp^2 \sim \mathcal{O}(\lambda^2 m_b^2) \; . \tag{4.136}$$

Since $(n_+p) \sim \mathcal{O}(m_b)$ we have to have $(n_+p) \sim \mathcal{O}(\lambda^2 m_b)$, while $p_\perp \sim \mathcal{O}(\lambda m)$, so the n_- component (n_+p) is down by one power of λ compared with the $p_\perp$ component. Thus we have to include two kinds of "soft" degrees of freedom, one of which scales as λ (which we shall call the soft degrees of freedom) and the other of which scales as λ^2 (which we shall call the ultrasoft degrees of freedom).

As discussed in the last section, the endpoint region of inclusive decays is defined by $p^2 \sim \Lambda_{\rm QCD} m_b$, which means that, in the case at hand, that λ scales as $\sqrt{\Lambda_{\rm QCD}/m_b}$. This specific to the power counting for inclusive decays; for exclusive channels, where SCET also applies the power counting has to be different (see below).

One important consequence is that the light degrees of freedom of a heavy hadron (given by the residual momentum k of the heavy quark) are actually ultrasoft degrees of freedom. Since we have $(n_+p) \sim \mathcal{O}(\lambda^2 m_b)$, we have to have $(n_+k) \sim \mathcal{O}(\lambda^2 m_b)$ as well; however, all momentum components of the residual momentum scale the same way, so we have $k \sim \mathcal{O}(\lambda^2 m_b)$, i.e. they are ultrasoft degrees of freedom.

Calculations in SCET are usually performed by constructing the Lagrangian for a collinear quark, from which Feynman rules have be derived. The derivation is in fact very similar to that for HQET discussed in some detail in Sect. 4.3. We may start out from the Lagrangian of a massless quark q,

$$\mathcal{L} = \bar{q} i \not{D} q \; , \tag{4.137}$$

where D denotes the covariant derivative of QCD. We shall discuss the dynamics of this quark under the above kinematical assumptions, and so we may use the two light-cone vectors to define the projectors

$$\mathcal{P} = \frac{1}{4} \not{n}_- \not{n}_+ \; , \qquad \mathcal{Q} = \frac{1}{4} \not{n}_+ \not{n}_- \; , \text{where} \quad \mathcal{P} + \mathcal{Q} = 1 \; . \tag{4.138}$$

In a similar way to that used in HQET, we can split the quark field q into two components

$$\xi = \mathcal{P} q \; , \qquad \eta = \mathcal{Q} q \; . \tag{4.139}$$

Likewise, we can decompose

$$\not{D} = \frac{1}{2} \not{n}_+ (in_- D) + \frac{1}{2} \not{n}_- (in_+ D) + \not{D}_\perp \; . \tag{4.140}$$

Inserting this and using $\not{n}_- \xi = 0 = \not{n}_+ \eta$, we obtain

$$\mathcal{L} = \frac{1}{2} \bar{\xi} \not{n}_+ (in_- D) \xi + \frac{1}{2} \bar{\eta} \not{n}_- (in_+ D) \eta + \bar{\xi} i \not{D}_\perp \eta + \bar{\eta} i \not{D}_\perp \xi \; . \tag{4.141}$$

The next step is to use the power counting defined above to identify the degrees of freedom that can be integrated out. Since $(in_+ D) \sim m$ and

$(in_- D) \sim \lambda^2 m$, we want to integrate out the η field. This can be done explicitly in the same way as was done in the HQET case by integrating over the small component field H_v (see (4.51)) in the Green's functions, written as a functional integral over the quark fields. As in HQET, this integration is a Gaussian integration, which can be done explicitly.

Performing this integration corresponds to using the equations of motion[9]

$$\eta = -\frac{1}{in_+ D + i\epsilon} \frac{\not{n}_+}{2} i \not{D}_\perp \xi \tag{4.142}$$

and inserting this result back into the Lagrangian

$$\mathcal{L} = \frac{1}{2} \bar{\xi} \not{n}_+ (in_- D) \xi - \bar{\xi} i \not{D}_\perp \frac{1}{in_+ D + i\epsilon} \frac{\not{n}_+}{2} i \not{D}_\perp \xi \, . \tag{4.143}$$

This resulting Lagrangian is still completely equivalent to that of full QCD, but it is now expressed in terms of the collinear quark field. However, it is non-local and becomes local only after expansion. To perform this expansion, we have to identify the large contribution in the quantity $in_+ D$. In order to do so, we split the gluon field A into a collinear contribution A_c and an ultrasoft contribution A_us

$$in_+ D = in_+ \partial + g n_+ A_\mathrm{c} + g n_+ A_\mathrm{us} = in_+ D_\mathrm{c} + g n_+ A_\mathrm{us} \tag{4.144}$$

where we have defined a collinear covariant derivative $iD_\mathrm{c} = i\partial + gA_\mathrm{c}$ containing the collinear gluon field. We now expand in the ultrasoft contribution, since it scales as $m\lambda^2$. This expansion corresponds to that of (4.53) in Sect. 4.3.

In order to have the complete Lagrangian, we need to do a similar decomposition for the gluonic part of the QCD Lagrangian. Furthermore, we also need to introduce ultrasoft quarks which appear, for example, as spectator quarks in a heavy hadron.

Note that in the leading-order Lagrangian, the only coupling to ultrasoft degrees of freedom is the coupling from $(in_- D)$ to the collinear quarks. A very similar coupling appears in the gluonic sector, where one has an $n_- A_\mathrm{us}$ coupling of ultrasoft gluons to collinear gluons. This observation is the basis for factorization theorems, which have been investigated intensively. In particular, one may derive a factorization statement for exclusive non-leptonic decays, which puts the naive factorization used for phenomenological estimates on a new basis. We shall not discuss this in detail here; rather, we discuss a particular example, which is again $B \to X_s \gamma$.

In order to discuss this example, we have to derive first the leading-order matching of a heavy-to-light current taking into account the kinematical situation described above. Naively one would match this current as as

[9] The Gaussian integration again also yields a determinant, which is, for the same reasons as in HQET, just an irrelevant constant.

$$\bar{Q}\Gamma q \longrightarrow \bar{h}_v \Gamma \xi \,, \tag{4.145}$$

since the light quark q turns into a collinear quark. However, the emission of a collinear gluon from the ingoing heavy quark will put this quark far off its mass shell and hence one would expect to obtain again a local interaction. This situation is depicted in Fig. 4.2 It has been shown in [64] that the emission of all these gluons can be resummed into a Wilson line $\mathcal{W}(0)$, such that the correct matching is

$$\bar{Q}\Gamma q = \bar{h}_v \mathcal{W}(0) \Gamma \xi \tag{4.146}$$

where

$$\mathcal{W}(x) = \mathrm{Pexp}\left(-i\int_{-\infty}^{0} ds\; n_- \cdot A_\mathrm{c}(x + sn_-)\right) \;; \tag{4.147}$$

here "Pexp" denotes the path-ordered exponential.

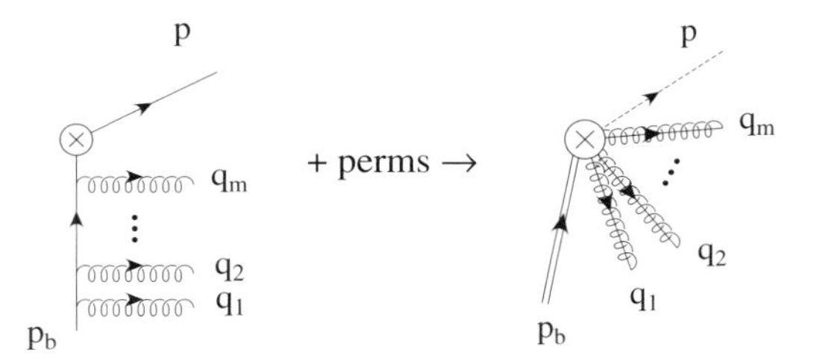

Fig. 4.2. Matching of the heavy-to-light current in SCET

In order to compute the rate, we consider again the correlator discussed in (4.101). Performing the matching to leading order, we obtain

$$\begin{aligned} R = &\int d^4x \exp(-ix[m_b v - q]) \\ &\times \langle B(p_B)|\bar{b}_v(0)\Gamma\mathcal{W}(0)\xi(0)\bar{\xi}(x)\mathcal{W}^\dagger(x)\Gamma^\dagger b_v(x)|B(p_B)\rangle \,. \end{aligned} \tag{4.148}$$

The key observation for obtaining factorization is that the ultrasoft degrees of freedom may be decoupled by a field redefinition. The procedure is very similar to what can be done in the context of the QED infrared problem [31], where the electron can be "dressed" with (ultra)soft photons by a field redefinition. For the case at hand, we can find a unitary transformation

$$Y(x) = \mathrm{Pexp}\left(-i\int_{-\infty}^{0} ds\; n_- \cdot A_\mathrm{us}(x + sn_-)\right) \tag{4.149}$$

which removes all ultrasoft interactions from the leading-order Lagrangian of the field ξ, since this depends only on the component $n_- \cdot A$ of the ultrasoft gluon field.

In terms of the "dressed" field

$$\xi^{(0)}(x) = Y(x)\xi(x) \,, \tag{4.150}$$

the leading-order Lagrangian

$$\mathcal{L}_{\rm l.o.} = \frac{1}{2}\bar{\xi}\not{n}_+(in_-D)\xi = \frac{1}{2}\bar{\xi}^{(0)}\not{n}_+(in_-D_c)\xi^{(0)} \tag{4.151}$$

does not depend on the ultrasoft gluon field any more, so any interaction with the ultrasoft degrees of freedom has disappeared from the leading order.

Since the B meson contains only ultrasoft degrees of freedom, the matrix element appearing in (4.148) can be factorized into a matrix element involving the B meson and one containing only collinear fields. We obtain

$$\begin{aligned} R = \int d^4x \exp(-ix[m_b v - q])\langle B(p_B)|\bar{b}_v(0)Y^\dagger(0)Y(x)b_v(x)|B(p_B)\rangle \\ \times \mathrm{Tr}\,\langle 0|\Gamma\mathcal{W}(0)\xi^{(0)}(0)\bar{\xi}^{(0)}(x)\mathcal{W}^\dagger(x)\Gamma^\dagger|0\rangle \,, \end{aligned} \tag{4.152}$$

where the trace refers to the Dirac indices.[10] Here we have explicitly written the "dressed" b quark field to show that the ultrasoft matrix element is simply a Wilson line connecting the two heavy quark fields at the points 0 and x.

We can rewrite (4.152) further, making use of the special kinematics of SCET. Fourier-transforming the matrix element involving the collinear quark fields,

$$\mathrm{Tr}\,\langle 0|\Gamma\mathcal{W}(0)\xi^{(0)}(0)\bar{\xi}^{(0)}(x)\mathcal{W}^\dagger(x)\Gamma^\dagger|0\rangle = \int \frac{d^4l}{(2\pi)^4} e^{-ilx}\hat{J}(l) \tag{4.153}$$

we find $l = mv + k - q$, where k is the residual momentum of the heavy quark (i.e. the conjugate variable of x). Using the SCET power counting (4.133) we find that $l_+ = m/2$ and hence we may neglect k_- in the momentum conservation. Likewise, $l_\perp$ is assumed to scale as λ (see (4.133)), so we can neglect $k_\perp$ in the momentum balance. Thus we end up having only k_+ in the momentum balance of the process, which in coordinate space corresponds to keeping only the dependence on $x_- = n_-x$ in the matrix element involving the B mesons,[11]

$$R = \int d^4x\, e^{-it(m_b - n_+q)}\langle B(v)|\bar{b}_v(0)Y^\dagger(0)Y(t)b_v(t)|B(v)\rangle J(t) \,, \tag{4.154}$$

where we have used the light-cone variable $t = n_-x/2$ and the "jet function"

[10] Note that we have made use of heavy-quark spin symmetry, ensuring that only the unit matrix can appear as the Dirac structure between the heavy-quark fields.

[11] It has been shown in [67] that this can be systematized in the form of a light-cone multipole expansion of the ultrasoft fields, which in the case at hand is the (static) b quark field.

$$J(t) = \int dx_+ \, d^2 x_\perp \, e^{\left(i \frac{m}{2} x_+\right)} \mathrm{Tr} \, \langle 0 | \Gamma \mathcal{W}(0) \xi^{(0)}(0) \bar{\xi}^{(0)}(x) \mathcal{W}^\dagger(x) \Gamma^\dagger | 0 \rangle \, . \quad (4.155)$$

This result, namely that in this particular kinematic regime the rates can be expressed in terms of a hard coefficient multiplied by a convolution of a jet function (containing the collinear contributions) and a shape function (containing the ultrasoft contributions) was derived in [69] even before SCET had been formulated.

An effective-field-theory calculation allows a systematic discussion of radiative corrections and a renormalization-group resummation of large logarithms induced by radiative corrections. One can even use the machinery of SCET to obtain the correct structure of the doubly logarithmic terms [70, 71].

SCET has also been extended to describe exclusive channels. However, the power counting needs to be modified, since the mass of an exclusive state of a light meson is now of order Λ_{QCD}, while for the inclusive channel considered above the mass has been of the order $\sqrt{m_b \Lambda_{\mathrm{QCD}}}$. This requires us also to introduce, in addition to the hard collinear modes considered up to now, new modes which have a different scaling from the scaling (4.133) [12] The SCET with the power counting appropriate to exclusive channels is sometimes called $\mathrm{SCET_{II}}$ as compared with $\mathrm{SCET_{I}}$ for inclusive channels.

One of the main results concerning exclusive non-leptonic channels is the proof of factorization for the decay mode $\overline{B}^0 \to D^+ \pi^-$ [73]. To leading order in the large mass expansion the decay amplitude factorizes into the $B \to \pi$ form factor and the pion decay constant; all corrections to this statement are either of the order $\alpha_s(\Lambda_{\mathrm{QCD}} m_b)$ (and thus perturbatively calculable) or suppressed by powers of $1/m_b$. We shall not go into any more details here, since the investigation of the properties of SCET is currently in progress. Some of the phenomenology of exclusive non-leptonic decays is discussed in Sect. 5.4.

The development of SCET was pre-dated by a method called QCD factorization [74, 75]. Although this method is not an obvious effective-field-theory approach (for example, it does not rely on the construction of an effective Lagrangian) it is still along similar lines, since it uses the heavy-mass limit and performs a systematic power counting. However, the power counting is performed on the basis of the Feynman diagrams of full QCD, and hence the proofs in QCD factorization are proofs valid up to a fixed order in QCD perturbation theory.

Exclusive non-leptonic decays into non-charmed final states have been investigated using QCD factorization also [76]. At one loop, it has been shown that also these decays factorize similarly to $\overline{B}^0 \to D^+ \pi^-$ to leading order in the expansion in $1/m_b$. Again, these issues are still under investigation and we shall not consider any more details here. In Sect. 5.4 we shall show some of the phenomenological results.

[12] Some material can be found in in [64, 65]; a more recent discussion can be found in [72].

4.8 Chiral Perturbation Theory

Another effective field theory which is often used in the context of weak decays is chiral perturbation theory. This effective theory describes QCD at very low energies and is mainly determined by the symmetries of QCD and the assumptions about how these symmetries are broken. It has been shown in [78] that this information is sufficient to fix the interactions of the Goldstone modes of the broken symmetry at low energies.

In a world with massless quarks (which is a good approximation for the light quarks, i.e for the up, down and strange quarks), all quark flavours and, separately, all quark helicities couple in the same way to the gluons. Technically, this means that for three (approximately) massless quarks, QCD exhibits a chiral $SU(3)_{\mathrm{L}} \times SU(3)_{\mathrm{R}}$ symmetry, where the subscripts stand for the left- and right-handed quarks.

If such a symmetry were realized in nature, one would observe parity doublets of $SU(3)$ multiplets of particles. While an $SU(3)$ flavour symmmetry is indeed observed (the well-known isospin being an $SU(2)$ subgroup of this flavour group $SU(3)$), there is no parity doubling, at least for the light hadrons such as the pion. Consequently one of the $SU(3)$ groups has to be broken. This breaking has to be spontaneous, since the Lagrangian clearly has the $SU(3)_L \times SU(3)_R$ symmetry.

One well-known consequence of spontaneous symmetry breaking is the appearance of massless modes [77], and in the case at hand one expects the appearance of an octet of massless particles corresponding to the spontaneous breaking of $SU(3)_{\mathrm{L}} \times SU(3)_{\mathrm{R}} \to SU(3)_{\mathrm{V}}$, where $SU(3)_{\mathrm{V}}$ is the vector subgoup of $SU(3)_{\mathrm{L}} \times SU(3)_{\mathrm{R}}$. Hence the axial-vector subgroup is broken and the corresponding Goldstone modes are pseudoscalar particles.

Thus this scenario predicts the presence of light pseudoscalar particles, which are identifed with the light pseudoscalar octet consisting of pions and kaons. These particles are not massless; however, their masses are small compared with the typical hadronic scale set by, for example, the proton mass or the mass of the ρ meson. Their masses may be related to the appearance of quark masses breaking the chiral $SU(3)_L \times SU(3)_R$ symmetry.

As a consequence of these symmetries, the currents

$$V_\mu^a = \bar{q}_i \gamma_\mu T_{ij}^a q_j \ , \quad \text{and} \quad A_\mu^a = \bar{q}_i \gamma_\mu \gamma_5 T_{ij}^a q_j \tag{4.156}$$

are conserved currents, since they generate the $SU(3)_L \times SU(3)_R \sim SU(3)_V \times SU(3)_A$ symmetry, where $q_1 = u$, $q_2 = d$, $q_3 = s$ and T^a are the generators of $SU(3)$ in the fundamental representation. While the currents V_μ^a generate the usual $SU(3)$ flavour symmetry, the axial currents generate the octet of massless states from the vacuum, such that

$$\langle 0 | A_\mu^a | \pi^b(p) \rangle = i f_\pi \delta^{ab} p_\mu \ , \tag{4.157}$$

where f_π is the pion decay constant which is determined by the decay of the charged pion $\pi \to \mu\bar{\nu}_\mu$. This process will be considered in some detail in Sect. 5.2.

The spontaneous breaking of the $SU(3)_\mathrm{L} \times SU(3)_\mathrm{R}$ symmetry down to $SU(3)_\mathrm{V}$ is signalled by the fact that the vacuum is not invariant any more under $SU(3)_\mathrm{L} \times SU(3)_\mathrm{R}$. The combination $\bar{q}q = (\bar{u}u + \bar{d}d + \bar{s}s)/3$ couples left- and right-handed components of the quarks and hence is not invariant under $SU(3)_\mathrm{L} \times SU(3)_\mathrm{R}$. A non-vanishing quark condensate

$$\langle 0|\bar{q}q|0\rangle \neq 0 \tag{4.158}$$

indicates that $SU(3)_L \times SU(3)_R$ is indeed spontaneously broken.

At very low energies, the complete dynamics of QCD is determined by the Goldstone modes π^a [78]. Using the currents as interpolating fields for pions (which is the main asumption of the "partially conserved axial vector current" (PCAC) hypothesis) one may compute transition matrix elements solely in terms of the algebra of the generators of the spontaneously broken $SU(3)_L \times SU(3)_R$ symmetry [79, 80].

Alternatively, since the dynamics of Goldstone modes is entirely fixed by the (broken) symmetry, one may equally well write down a Lagrangian encoding this information. Starting from the $SU(3)_\mathrm{L} \times SU(3)_\mathrm{R}$ symmetry, we may define a field Σ with the transformation property

$$\Sigma \to L\Sigma R^\dagger \,, \quad \text{for} \quad L \in SU(3)_\mathrm{L} \text{ and } R \in SU(3)_\mathrm{R} \,, \tag{4.159}$$

and the simplest $SU(3)_\mathrm{L} \times SU(3)_\mathrm{R}$ invariant Lagrangian is given by

$$\mathcal{L} = \frac{f^2}{4} \mathrm{Tr}\left[(\partial_\mu \Sigma)^\dagger (\partial_\mu \Sigma)\right] \,, \tag{4.160}$$

where f is a constant which is considered below. Note that (4.160) is the Lagrangian of the non-linear σ model.

The $SU(3)_L \times SU(3)_R$ symmetry is broken by a vacuum expectation value of Σ such that

$$\langle 0|\Sigma|0\rangle = 1 \,. \tag{4.161}$$

This breaks the $SU(3)_\mathrm{L} \times SU(3)_\mathrm{R}$ symmetry down to $SU(3)_\mathrm{L+R} = SU(3)_\mathrm{V}$ which means that the transformations L and R appearing in (4.159) have to be equal, i.e. $L = R$.

Owing to this breaking Goldstone bosons have to appear and the field Σ can be expressed in terms of these Goldstone modes. One possible representation is given by

$$\Sigma = \exp\left(\frac{i}{f} T^a \pi^a\right) = \begin{bmatrix} \pi_0/\sqrt{2} + \eta_8/\sqrt{6} & \pi^+ & K^+ \\ \pi^- & -\pi_0/\sqrt{2} + \eta_8/\sqrt{6} & K^0 \\ K^- & \overline{K}^0 & -2\eta_8/\sqrt{6} \end{bmatrix} \tag{4.162}$$

where η_8 is the octet contribution of the η, which is, owing to η–η' mixing, a linear combination of the physical η and η' mesons.

From (4.160) and (4.162) we can obtain the possible interactions between the Goldstone bosons. In particular, we may compute the currents V_μ and A_μ from Noether's theorem, and we find

$$\begin{aligned} V_\mu^a &= \frac{f^2}{4}\mathrm{Tr}\left[\Sigma T^a(\partial_\mu \Sigma)^\dagger + \Sigma^\dagger T^a(\partial_\mu \Sigma)\right] \\ &= i f^{abc}[\pi^b\,,\,(\partial_\mu \pi^c)] + \cdots\,, \end{aligned} \tag{4.163}$$

$$A_\mu^a = \frac{f^2}{4}\mathrm{Tr}\left[\Sigma T^a(\partial_\mu \Sigma)^\dagger - \Sigma^\dagger T^a(\partial_\mu \Sigma)\right] = i f(\partial_\mu \pi^a) + \cdots\,, \tag{4.164}$$

where the ellipses denote terms containing higher powers of fields. Comparing (4.164) with (4.157), we infer that the normalization constant f appearing in the Lagrangian is in fact the pion decay constant f_π.

The difference between the chiral Lagrangian (4.160) and the other effective field theories discussed so far is that the matching of the chiral Lagrangian to full QCD cannot be performed explicitly. In principle, f_π is related to the scale parameter Λ_{QCD} of QCD, but the matching cannot be done, since all the matching procedures discussed so far are perturbative.

Still (4.160) is a good starting point for a systematic expansion, which is called chiral perturbation theory [81, 82]. The Lagrangian (4.160) is valid for small momenta of the Goldstone bosons, where these bosons are the only degrees of freedom present. This also indicates the region of validity, since at scales of the order of the ρ-meson mass (which are the first "non-Goldstone" excitations of a quark–anti quark pair), one expects to have additional interactions which are not determined by the Goldstone modes alone. This scale is usually called the chiral-symmetry-breaking scale $\Lambda_{\chi SB}$ and is, by the rules of "naive dimenional analysis", [1, 83, 84] of the order of $4\pi f_\pi$.

The idea of chiral perturbation theory is to perform a systematic expansion in the ratio $p_\pi/\Lambda_{\chi SB}$, where p_π is the typical momentum of the Goldstone boson. This means that the subleading terms that need to be added to (4.160) are the ones that contain more derivatives acting on the Goldstone boson field Σ and which are compatible with the (spontaneously broken) $SU(3)_\mathrm{L} \times SU(3)_\mathrm{R}$ symmetry.

Once subleading terms are considered, one also has to take into account the presence of the *explicit* breaking of the $SU(3)_L \times SU(3)_R$ symmetry through the small quark masses, since the masses are usually counted in the same way as the momenta of the mesons. The breaking by quark mass terms can be discussed by introducing a spurion field $\mathcal{M}$, for which we assume the transformation property

$$\mathcal{M} \to L\mathcal{M}R^\dagger\,, \text{ for } \quad L \in SU(3)_\mathrm{L} \text{ and } R \in SU(3)_\mathrm{R}\,. \tag{4.165}$$

The Lagrangian is supplemented by adding all $SU(3)_\mathrm{L} \times SU(3)_\mathrm{R}$-invariant terms involving the spurion field, where one power of the spurion counts in the same way as one power of the momentum. Symmetry breaking is now implemented by replacing the spurion field by the fixed matrix

$$\mathcal{M} = \begin{pmatrix} m_u & 0 & 0 \\ 0 & m_d & 0 \\ 0 & 0 & m_s \end{pmatrix} . \qquad (4.166)$$

The simplest (i.e. of lowest order in the chiral expansion) term one can write is

$$\mathcal{L}_{\mathrm{m}} = \frac{f^2 B_0}{2} \mathrm{Tr} \left[\mathcal{M}^\dagger \Sigma + \Sigma^\dagger \mathcal{M} \right] , \qquad (4.167)$$

which leads to a mass term for the Goldstone bosons. Here B_0 is a new constant.

The peculiar feature of the mass term (4.167) is that it implies that the the square of the mass of the pseudo-Goldstone particles is linear in the quark masses [85]. In fact, when we work out the details of these relations, we obtain

$$\begin{aligned} m_{\pi^\pm}^2 &= B_0(m_u + m_d), & m_{K^\pm}^2 &= B_0(m_u + m_s), \\ m_{K^0}^2 &= B_0(m_s + m_d), & m_{\eta_8}^2 &= \frac{B_0}{3}(m_u + m_d + 4m_s) . \end{aligned} \qquad (4.168)$$

Eliminating the quark masses from these relations, we obtain the well-known Gell–Mann Okubo mass relations [86, 87], which have been obtained solely from $SU(3)$ arguments. In fact, on the basis of symmetry only, we could also have obtained similar relations relating the quark masses rather than their squares, but the chiral Lagrangian predicts that the relations should hold for the squares of the masses, which is in accordance with observations. Furthermore, these relations predict the correct mass for the η meson.

Finally, the parameter B_0 can be related to the quark condensate using soft-pion techniques [81]; one finds

$$B_0 = -\frac{1}{f_\pi^2} \langle 0 | \bar{q} q | 0 \rangle , \qquad (4.169)$$

where consistency requires that the quark condensate has to be negative.

References

1. S. Weinberg, Physica A **96**, 327 (1979).
2. J. Polchinski, Nucl. Phys. B **231**, 269 (1984).
3. H. Georgi, Annu. Rev. Nucl. Part. Sci. **43**, 209 (1993).
4. H. M. Georgi, in *The New Physics*, ed. P. Davies, p. (...).
5. A. V. Manohar, in *Perturbative and nonperturbative aspects of quantum field theory*, pp. 311-362, lectures at the 35th Internationale Universitätswochen fuer Kern- und Teilchenphysik, Schladming, Austria, 2-9 Mar 1996 [arXiv:hep-ph/9606222].

6. T. Appelquist and J. Carazzone, Phys. Rev. D **11**, 2856 (1975).
7. R. D. C. Miller, Phys. Rep. **106** (1984) 169.
8. K. G. Wilson and J. B. Kogut, Phys. Rep. **12**, 75 (1974).
9. J. Collins, *Renormalization*, Cambridge Monographs on Mathematical Physics (Cambridge University Press, Cambridge, 1987).
10. S. Weinberg, Phys. Rev. D **8**, 3497 (1973).
11. G. Buchalla, A. J. Buras and M. E. Lautenbacher, Rev. Mod. Phys. **68**, 1125 (1996) [arXiv:hep-ph/9512380].
12. W. Kilian, *Electroweak Symmetry Breaking: The Bottom-Up Approach*, Springer tracts in modern physics. No. 198 (Springer, Berlin, Heidelberg, 2003).
13. L. D. Faddeev and A. A. Slavnov, *Gauge Fields. Introduction To Quantum Theory*, Frontiers in Physics, No. 83, (Addison-Wesley, Redwood City,1990).
14. C. Quigg, *Gauge Theories of the Strong, Weak and Electromagnetic Interactions*, Frontiers in Physics, No. 56, (Addison-Wesley, Redwood City, 1983).
15. K. Huang, *Quarks Leptons and Gauge Fields* (World Scientific, Singapore, 1992).
16. P. Renton, *Electroweak Interactions* (Cambridge University Press, Cambridge, 1990)
17. J. Donoghue, E. Golowich, B. Holstein, *Dynamics of the Standard Model*, Cambridge Monographs on Particle Physics (Cambridge University Press, Cambridge, 1992)
18. O. Nachtmann, *Elementary Particle Physics*, Springer Texts and Monographs in Physics (Springer, Berlin, Heidelberg, 1989).
19. M. Böhm, A. Denner and H. Joos, *Gauge Theories of the Strong and Electroweak Interaction* (Teubner, Stuttgart, 2001)
20. N. Isgur and M. B. Wise, Phys. Lett. B **232**, 113 (1989).
21. N. Isgur and M. B. Wise, Phys. Lett. B **237**, 527 (1990).
22. M. A. Shifman and M. B. Voloshin, Sov. J. Nucl. Phys. **47**, 511 (1988) [Yad. Fiz. **47**, 801 (1988)].
23. B. Grinstein, Nucl. Phys. B **339**, 253 (1990).
24. H. Georgi, Phys. Lett. B **240**, 447 (1990).
25. A. F. Falk, H. Georgi, B. Grinstein and M. B. Wise, Nucl. Phys. B **343**, 1 (1990).
26. M. Neubert, Phys. Rep. **245**, 259 (1994) [arXiv:hep-ph/9306320].
27. T. Mannel, Rept. Prog. Phys. **60**, 1113 (1997).
28. I. I. Y. Bigi, M. A. Shifman and N. Uraltsev, Annu. Rev. Nucl. Part. Sci. **47**, 591 (1997) [arXiv:hep-ph/9703290].
29. A. V. Manohar and M. B. Wise, *Heavy Quark Physics*, Cambridge Monographs on Particle Physics, Nuclear Physics and Cosmology, No. 10 (Cambridge University Press, Cambridge, 2000).
30. E. Eichten and B. Hill, Phys. Lett. B **234**, 511 (1990).
31. F. Bloch and A. Nordsieck, Phys. Rev. **52**, 54 (1937).
32. T. Mannel, W. Roberts and Z. Ryzak, Nucl. Phys. B **368**, 204 (1992).
33. E. Uehling, Phys. Rev. **48**, 55 (1935).
34. F. Foldy and S. Wouthuysen, Phys. Rev. **78**, 29 (1950).
35. S. Balk, J. G. Korner and D. Pirjol, Nucl. Phys. B **428**, 499 (1994) [arXiv:hep-ph/9307230].
36. M. E. Luke, Phys. Lett. B **252**, 447 (1990).

37. M. Ademollo and R. Gatto, Phys. Rev. Lett. **13**, 264 (1964).
38. M. Luke and A. Manohar, Phys. Lett. B **286**, 348 (1992).
39. Y. Chen, Phys. Lett. **B317**, 421 (1993).
40. I. I. Y. Bigi, N. G. Uraltsev and A. I. Vainshtein, Phys. Lett. B **293**, 430 (1992); erratum, Phys. Lett. B **297**, 477 (1993) [arXiv:hep-ph/9207214].
41. I. I. Y. Bigi, M. A. Shifman, N. G. Uraltsev and A. I. Vainshtein, Phys. Rev. Lett. **71**, 496 (1993). [arXiv:hep-ph/9304225].
42. I. I. Y. Bigi, B. Blok, M. A. Shifman and A. I. Vainshtein, Phys. Lett. B **323** (1994) 408 [arXiv:hep-ph/9311339].
43. J. Chay, H. Georgi and B. Grinstein, Phys. Lett. B **247**, 399 (1990).
44. A. V. Manohar and M. B. Wise, Phys. Rev. D **49** 1310 (1994) [arXiv:hep-ph/9308246].
45. T. Mannel, Nucl. Phys. B **413**, 396 (1994) [arXiv:hep-ph/9308262].
46. B. Guberina, S. Nussinov, R. D. Peccei and R. Ruckl, Phys. Lett. B **89**, 111 (1979).
47. B. Guberina, R. D. Peccei and R. Ruckl, Phys. Lett. B **90**, 169 (1980).
48. B. Guberina, R. D. Peccei and R. Ruckl, Phys. Lett. B **91**, 116 (1980).
49. I. I. Y. Bigi and N. Uraltsev, Int. J. Mod. Phys. A **16**, 5201 (2001) [arXiv:hep-ph/0106346].
50. M. A. Shifman, in *At the Frontier of Particle Physics*, ed. M. Shifman, Boris Ioffe Festschrift, p. 1447 (World Scientific, Singapore, 2001). arXiv:hep-ph/0009131.
51. I. I. Bigi and T. Mannel, contribution to workshop on the CKM matrix and the unitarity triangle, Geneva, Switzerland, 13–16 Feb. 2002, arXiv:hep-ph/0212021.
52. I. I. Y. Bigi, M. A. Shifman, N. G. Uraltsev and A. I. Vainshtein, Phys. Rev. D **52**, 196 (1995) [arXiv:hep-ph/9405410].
53. T. Mannel, Phys. Rev. D **50**, 428 (1994) [arXiv:hep-ph/9403249].
54. M. Neubert, Phys. Rev. D **49**, 3392 (1994) [arXiv:hep-ph/9311325].
55. M. Neubert, Phys. Rev. D **49**, 4623 (1994) [arXiv:hep-ph/9312311].
56. I. I. Y. Bigi, M. A. Shifman, N. G. Uraltsev and A. I. Vainshtein, Int. J. Mod. Phys. A **9**, 2467 (1994) [arXiv:hep-ph/9312359].
57. T. Mannel and M. Neubert, Phys. Rev. D **50**, 2037 (1994) [arXiv:hep-ph/9402288].
58. C. W. Bauer, M. E. Luke and T. Mannel, Phys. Rev. D **68**, 094001 (2003) [arXiv:hep-ph/0102089].
59. C. W. Bauer, M. Luke and T. Mannel, Phys. Lett. B **543**, 261 (2002) [arXiv:hep-ph/0205150].
60. A. K. Leibovich, Z. Ligeti and M. B. Wise, Phys. Lett. B **539**, 242 (2002) [arXiv:hep-ph/0205148].
61. M. Neubert, Phys. Lett. B **543**, 269 (2002) [arXiv:hep-ph/0207002].
62. F. Campanario and T. Mannel, Phys. Rev. D **65**, 094017 (2002) [arXiv:hep-ph/0201136].
63. A. V. Manohar, T. Mehen, D. Pirjol and I. W. Stewart, Phys. Lett. B **539**, 59 (2002) [arXiv:hep-ph/0204229].
64. C. W. Bauer, S. Fleming, D. Pirjol and I. W. Stewart, Phys. Rev. D **63** (2001) 114020 [arXiv:hep-ph/0011336].

65. C. W. Bauer, D. Pirjol and I. W. Stewart, Phys. Rev. D **65**, 054022 (2002) [arXiv:hep-ph/0109045].
66. M. Beneke, A. P. Chapovsky, M. Diehl and T. Feldmann, Nucl. Phys. B **643**, 431 (2002) [arXiv:hep-ph/0206152].
67. M. Beneke and T. Feldmann, Phys. Lett. B **553**, 267 (2003) [arXiv:hep-ph/0211358].
68. M. J. Dugan and B. Grinstein, Phys. Lett. B **255**, 583 (1991).
69. G. P. Korchemsky and G. Sterman, Phys. Lett. B **340**, 96 (1994) [arXiv:hep-ph/9407344].
70. C. W. Bauer, S. Fleming and M. E. Luke, Phys. Rev. D **63**, 014006 (2001) [arXiv:hep-ph/0005275].
71. A. K. Leibovich, I. Low and I. Z. Rothstein, Phys. Lett. B **513**, 83 (2001) [arXiv:hep-ph/0105066].
72. T. Becher, R. J. Hill and M. Neubert, Phys. Rev. D **69**, 054017 (2004) [arXiv:hep-ph/0308122].
73. C. W. Bauer, D. Pirjol and I. W. Stewart, Phys. Rev. Lett. **87**, 201806 (2001) [arXiv:hep-ph/0107002].
74. M. Beneke, G. Buchalla, M. Neubert and C. T. Sachrajda, Phys. Rev. Lett. **83**, 1914 (1999) [arXiv:hep-ph/9905312].
75. M. Beneke, G. Buchalla, M. Neubert and C. T. Sachrajda, Nucl. Phys. B **591**, 313 (2000) [arXiv:hep-ph/0006124].
76. M. Beneke, G. Buchalla, M. Neubert and C. T. Sachrajda, Nucl. Phys. B **606**, 245 (2001) [arXiv:hep-ph/0104110].
77. J. Goldstone, Nuovo Cim. **19**, 154 (1961).
78. S. R. Coleman, J. Wess and B. Zumino, Phys. Rev. **177**, 2239 (1969).
79. S. Adler and R. Dashen, *Current Algebras* (Benjamin, New York 1968).
80. V. de Alfaro, S. Fubini, G. Furlan and C. Rosetti: *Currents in Hadron Physics* (North-Holland, Amsterdam 1973).
81. J. Gasser and H. Leutwyler, Nucl. Phys. B **250**, 465 (1985).
82. J. Gasser and H. Leutwyler, Ann. Phys. **158**, 142 (1984).
83. A. Manohar and H. Georgi, Nucl. Phys. **B234**, 189 (1984).
84. H. Georgi and L. Randall, Nucl. Phys. **B276**, 241 (1986).
85. M. Gell-Mann, R. Oakes and B. Renner, Phys. Rev. **175**, 2195 (1968).
86. M. Gell-Mann, Phys. Rev. **125**, 1067 (1962).
87. S. Okubo, Progr. Theor. Phys. **27**, 949 (1962).

5 Applications I: $\Delta F = 1$ Processes

In this chapter and the following one, we shall give a few sample applications of effective field theories to the flavour physics of the Standard Model. We shall discuss mainly the effective Hamiltonian for flavour transitions derived in the context of the Standard Model, but an extension to models beyond the Standard Model will be obvious. We shall consider first the Hamiltonian for $\Delta F = \pm 1$ processes, i.e. processes in which the flavour quantum numbers change by one unit; $\Delta F = \pm 2$ processes will be discussed in the next chapter. After a few remarks on light-hadron decays, we shall study bottom and charm decays, where one may use the heavy-quark mass expansion. We are not aiming at completeness; rather, we shall only demonstrate how effective-field-theory methods can be applied efficiently.

5.1 $\Delta F = 1$ Effective Hamiltonian

In this section we shall discuss the effective Hamiltonian relevant to decays of bottom, charm and strange hadrons. The masses of these particles are much smaller than the masses of the weak gauge bosons and of the top quark, and hence we can switch to an effective-theory description as discussed in Sect. 4.2.

Since the top quark and the weak bosons have masses of the same order, we may integrate out these particles at the same scale, which we choose to be the scale M_W. In the following subsections, we shall collect together the relevant formulae for the various weak transitions of quarks.

5.1.1 Effective Hamiltonian for Semileptonic Processes

Semileptonic processes are mediated by operators involving one hadronic and one leptonic current. Integrating out the weak bosons and the top quark yields at tree level, the effective Hamiltonian

$$\mathcal{H}_{eff}^{(sl)} = \frac{4G_F}{\sqrt{2}} \left(\bar{\mathcal{U}}_L \gamma_\mu V_{CKM} \mathcal{D}_L\right) \left(\bar{e}_L \gamma_\mu \bar{\nu}_{e,\mathrm{L}} + \bar{\mu}_L \gamma_\mu \bar{\nu}_{\mu,\mathrm{L}} + \bar{\tau}_L \gamma_\mu \bar{\nu}_{\tau,\mathrm{L}}\right) + \text{h.c.}\,, \tag{5.1}$$

where any operator involving the top quark is simply omitted and the matrix notation introduced earlier is understood. Writing explicitly the various possible transitions we obtain the following ($\ell = e, \mu$ or τ):

- Semileptonic decays of bottom hadrons:

$$\mathcal{H}_{eff}(b \to c\ell\bar{\nu}_\ell) = \frac{4G_F}{\sqrt{2}} V_{cb} \left(\bar{c}_{\rm L}\gamma_\mu b_{\rm L}\right) \left(\bar{\ell}_{\rm L}\gamma_\mu \nu_{\ell,{\rm L}}\right) + \text{h.c.}\,, \tag{5.2}$$

$$\mathcal{H}_{eff}(b \to u\ell\bar{\nu}_\ell) = \frac{4G_F}{\sqrt{2}} V_{ub} \left(\bar{u}_{\rm L}\gamma_\mu b_{\rm L}\right) \left(\bar{\ell}_{\rm L}\gamma_\mu \nu_{\ell,{\rm L}}\right) + \text{h.c.} \tag{5.3}$$

- Semileptonic kaon or hyperon decays:

$$\mathcal{H}_{eff}(s \to u\ell\bar{\nu}_\ell) = \frac{4G_F}{\sqrt{2}} V_{us} \left(\bar{u}_{\rm L}\gamma_\mu s_{\rm L}\right) \left(\bar{\ell}_{\rm L}\gamma_\mu \nu_{\ell,{\rm L}}\right) + \text{h.c.} \tag{5.4}$$

- Semileptonic decays of charmed hadrons:

$$\mathcal{H}_{eff}(c \to s\ell\bar{\nu}_\ell) = \frac{4G_F}{\sqrt{2}} V_{cs} \left(\bar{c}_{\rm L}\gamma_\mu s_{\rm L}\right) \left(\bar{\ell}_{\rm L}\gamma_\mu \nu_{\ell,{\rm L}}\right) + \text{h.c.}\,, \tag{5.5}$$

$$\mathcal{H}_{eff}(c \to d\ell\bar{\nu}_\ell) = \frac{4G_F}{\sqrt{2}} V_{cd} \left(\bar{c}_{\rm L}\gamma_\mu d_{\rm L}\right) \left(\bar{\ell}_{\rm L}\gamma_\mu \nu_{\ell,{\rm L}}\right) + \text{h.c.} \tag{5.6}$$

- β decays:

$$\mathcal{H}_{eff}(d \to u\ell\bar{\nu}_\ell) = \frac{4G_F}{\sqrt{2}} V_{ud} \left(\bar{u}_{\rm L}\gamma_\mu d_{\rm L}\right) \left(\bar{\ell}_{\rm L}\gamma_\mu \nu_{\ell,{\rm L}}\right) + \text{h.c.} \tag{5.7}$$

These results hold at tree level, i.e. at leading order in the strong coupling α_s. When (5.2)-(5.7) are interpreted as local operators multiplied by Wilson coefficients, these coefficients (after factoring out the common factor $G_F/\sqrt{2}$ and the CKM factors) are simply unity. Furthermore, since in the limit of vanishing quark masses both the axial-vector and the vector current are conserved, the operators appearing in the effective Hamiltonians for semileptonic transitions do not have an anomalous dimension in full QCD, owing to current conservation. This means that no large logarithms of the form $(\alpha_s/\pi)\ln(M_W^2/\mu^2)$ can appear.

5.1.2 Effective Hamiltonian for Non-Leptonic Processes

Non-leptonic processes are mediated by operators involving two hadronic currents. In the same way as for the semileptonic case, we obtain at tree level

$$\mathcal{H}_{eff}^{(sl)} = \frac{4G_F}{\sqrt{2}} \left(\bar{\mathcal{U}}_L\gamma_\mu V_{CKM}\mathcal{D}_L\right) \left(\bar{\mathcal{D}}_L\gamma_\mu V_{CKM}^\dagger \mathcal{U}_L\right) + \text{h.c.} \tag{5.8}$$

Similarly to the case of semileptonic decays we can decompose (5.8) into the various channels. We obtain the following:

- Non-leptonic decays of bottom hadrons:

$$\mathcal{H}_{eff}(b \to c\bar{u}d) = \frac{4G_F}{\sqrt{2}} V_{cb} V_{ud}^* \left(\bar{c}_{\rm L}\gamma_\mu b_{\rm L}\right) \left(\bar{d}_{\rm L}\gamma_\mu u_{\rm L}\right) + \text{h.c.} \,, \tag{5.9}$$

$$\mathcal{H}_{eff}(b \to c\bar{c}s) = \frac{4G_F}{\sqrt{2}} V_{cb} V_{cs}^* \left(\bar{c}_{\rm L}\gamma_\mu b_{\rm L}\right) \left(\bar{s}_{\rm L}\gamma_\mu c_{\rm L}\right) + \text{h.c.} \,, \tag{5.10}$$

$$\mathcal{H}_{eff}(b \to c\bar{u}s) = \frac{4G_F}{\sqrt{2}} V_{cb} V_{us}^* \left(\bar{c}_{\rm L}\gamma_\mu b_{\rm L}\right) \left(\bar{s}_{\rm L}\gamma_\mu u_{\rm L}\right) + \text{h.c.} \,, \tag{5.11}$$

$$\mathcal{H}_{eff}(b \to c\bar{c}d) = \frac{4G_F}{\sqrt{2}} V_{cb} V_{cd}^* \left(\bar{c}_{\rm L}\gamma_\mu b_{\rm L}\right) \left(\bar{d}_{\rm L}\gamma_\mu c_{\rm L}\right) + \text{h.c.} \,, \tag{5.12}$$

$$\mathcal{H}_{eff}(b \to u\bar{u}d) = \frac{4G_F}{\sqrt{2}} V_{ub} V_{ud}^* \left(\bar{u}_{\rm L}\gamma_\mu b_{\rm L}\right) \left(\bar{d}_{\rm L}\gamma_\mu u_{\rm L}\right) + \text{h.c.} \,, \tag{5.13}$$

$$\mathcal{H}_{eff}(b \to u\bar{c}s) = \frac{4G_F}{\sqrt{2}} V_{ub} V_{cs}^* \left(\bar{u}_{\rm L}\gamma_\mu b_{\rm L}\right) \left(\bar{s}_{\rm L}\gamma_\mu c_{\rm L}\right) + \text{h.c.} \,, \tag{5.14}$$

$$\mathcal{H}_{eff}(b \to u\bar{u}s) = \frac{4G_F}{\sqrt{2}} V_{ub} V_{us}^* \left(\bar{u}_{\rm L}\gamma_\mu b_{\rm L}\right) \left(\bar{s}_{\rm L}\gamma_\mu u_{\rm L}\right) + \text{h.c.} \,, \tag{5.15}$$

$$\mathcal{H}_{eff}(b \to u\bar{c}d) = \frac{4G_F}{\sqrt{2}} V_{ub} V_{cd}^* \left(\bar{u}_{\rm L}\gamma_\mu b_{\rm L}\right) \left(\bar{d}_{\rm L}\gamma_\mu c_{\rm L}\right) + \text{h.c.} \tag{5.16}$$

- Non-leptonic decays of charmed hadrons:

$$\mathcal{H}_{eff}(c \to su\bar{d}) = \frac{4G_F}{\sqrt{2}} V_{cs} V_{ud}^* \left(\bar{c}_{\rm L}\gamma_\mu s_{\rm L}\right) \left(\bar{d}_{\rm L}\gamma_\mu u_{\rm L}\right) + \text{h.c.} \,, \tag{5.17}$$

$$\mathcal{H}_{eff}(c \to su\bar{s}) = \frac{4G_F}{\sqrt{2}} V_{cs} V_{us}^* \left(\bar{c}_{\rm L}\gamma_\mu s_{\rm L}\right) \left(\bar{s}_{\rm L}\gamma_\mu u_{\rm L}\right) + \text{h.c.} \,, \tag{5.18}$$

$$\mathcal{H}_{eff}(c \to du\bar{d}) = \frac{4G_F}{\sqrt{2}} V_{cd} V_{ud}^* \left(\bar{c}_{\rm L}\gamma_\mu d_{\rm L}\right) \left(\bar{d}_{\rm L}\gamma_\mu u_{\rm L}\right) + \text{h.c.} \,, \tag{5.19}$$

$$\mathcal{H}_{eff}(c \to du\bar{s}) = \frac{4G_F}{\sqrt{2}} V_{cd} V_{us}^* \left(\bar{c}_{\rm L}\gamma_\mu d_{\rm L}\right) \left(\bar{s}_{\rm L}\gamma_\mu u_{\rm L}\right) + \text{h.c.} \tag{5.20}$$

- Non-leptonic decays of strange hadrons:

$$\mathcal{H}_{eff}(s \to u\bar{u}d) = \frac{4G_F}{\sqrt{2}} V_{us} V_{ud}^* \left(\bar{u}_{\rm L}\gamma_\mu s_{\rm L}\right) \left(\bar{d}_{\rm L}\gamma_\mu u_{\rm L}\right) + \text{h.c.} \tag{5.21}$$

These results have to be interpreted as the tree-level matching for the relevant operators that the effective Hamiltonian consists of. The current–current operators appearing in the effective Hamiltonian are dimension-six operators and, as discussed earlier, the renormalization of these operators induces mixing with other operators of dimension six.

The simplest case is when four different flavours are involved. As an example, we study the transition $b \to c\bar{u}d$. Owing to the fact that all quarks have to be left-handed, we can have only two dimension-six operators differing by their colour structure,

$$O_1 = (\bar{c}_{L,i}\gamma_\mu b_{L,j})\left(\bar{d}_{L,j}\gamma_\mu u_{L,i}\right) , \tag{5.22}$$
$$O_2 = (\bar{c}_{L,i}\gamma_\mu b_{L,i})\left(\bar{d}_{L,j}\gamma_\mu u_{L,j}\right) , \tag{5.23}$$

where $i, j = 1, 2, 3$ are the colour indices of the quarks. The effective Hamiltonian is then given by

$$\mathcal{H}_{eff}(b \to c\bar{u}d) = \frac{4G_F}{\sqrt{2}} V_{cb} V_{ud}^* \left[C_1(\mu)O_1(\mu) + C_2(\mu)O_2(\mu)\right] + \text{h.c.} , \tag{5.24}$$

where the coefficients C_1 and C_2 are determined from the renormalization group equations as discussed in Sect. 4.1. These have been calculated in [1] to leading logarithmic accuracy, and a detailed discussion of the next-to-leading logarithms can be found in [2].

The calculation of the anomalous-dimension matrix involves the diagrams depicted in Fig. 5.1. It is clear from the colour flow in the diagram, that the two operators O_1 and O_2 will mix under renormalization. The calculation of the one-loop anomalous dimension is straightforward, and one finds [1]

$$\gamma = \frac{\alpha_s}{4\pi}\begin{pmatrix} -2 & 6 \\ 6 & -2 \end{pmatrix} + \mathcal{O}(\alpha_s^2) . \tag{5.25}$$

Using this result together with the leading order term of the β function given in (4.17) and (4.18) in Sect. 4.1 we can solve the renormalization group (4.15) with the initial condition

$$C_1(M_W) = 0 \qquad \text{and} \qquad C_2(M_W) = 1 , \tag{5.26}$$

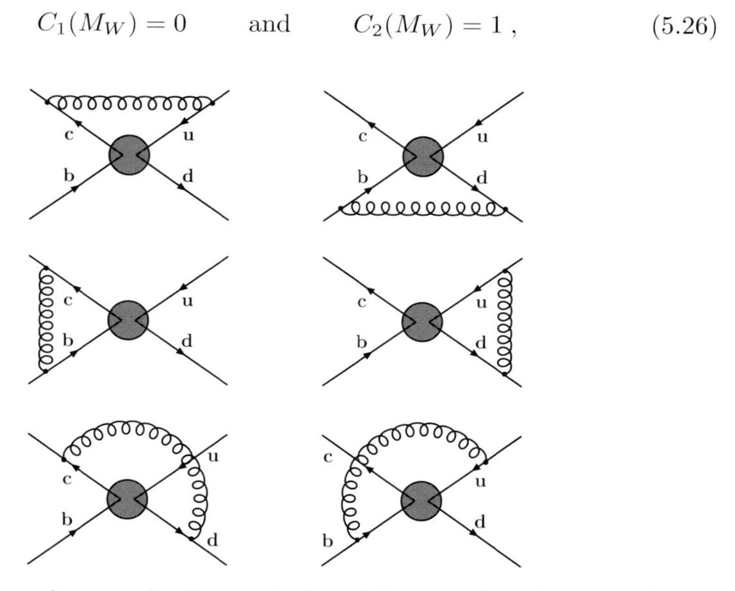

Fig. 5.1. Feynman diagrams for the calculation of the anomalous dimension (5.25). Self-energy diagrams are omitted

and obtain, for scales above the b quark mass,

$$C_1(\mu) = \frac{1}{2}\left[\left(\frac{\alpha_s(\mu)}{\alpha_s(M_W)}\right)^{-6/23} - \left(\frac{\alpha_s(\mu)}{\alpha_s(M_W)}\right)^{12/23}\right], \tag{5.27}$$

$$C_2(\mu) = \frac{1}{2}\left[\left(\frac{\alpha_s(\mu)}{\alpha_s(M_W)}\right)^{-6/23} + \left(\frac{\alpha_s(\mu)}{\alpha_s(M_W)}\right)^{12/23}\right]. \tag{5.28}$$

The typical scale for the matrix elements is the b quark mass, and hence we evolve the coefficients down to m_b. At m_b, the values of the coefficients are

$$C_1(m_b \sim 5\,\text{GeV}) \approx -0.25 \quad C_2(m_b \sim 5\,\text{GeV}) \approx 1.1\,, \tag{5.29}$$

indicating that the QCD corrections are sizeable.

We can discuss the transitions $b \to c\bar{u}s$, $b \to c\bar{u}s$, $b \to u\bar{c}s$, and $b \to u\bar{c}d$ in the same way, with the corresponding replacements in the operators O_1 and O_2.

Likewise, we can consider charm decays for a transition in which four different flavours are involved, i.e. for the transitions $c \to s\bar{d}u$ and $c \to d\bar{s}u$. As an example, we consider the case $c \to s\bar{d}u$ in which case the relevant operators are

$$\hat{O}_1 = (\bar{c}_{L,i}\gamma_\mu s_{L,j})\left(\bar{d}_{L,j}\gamma_\mu u_{L,i}\right), \tag{5.30}$$

$$\hat{O}_2 = (\bar{c}_{L,i}\gamma_\mu s_{L,i})\left(\bar{d}_{L,j}\gamma_\mu u_{L,j}\right). \tag{5.31}$$

If we evolve the coefficient down from the scale M_W, we have the same renormalization group equation as before. However, in this case one would like to evolve them down to a scale $\mu \sim m_c$ and so we have to take into account the fact that at the scale of the b quark mass we have a change from a theory with five active flavours to a theory with four active flavours. The operator and the anomalous dimensions (at one-loop level) remain the same, except that the number of active flavours changes from five to four in the β function when we pass through the point $\mu = m_b$. Taking this into account we obtain the following for the coefficients at $\mu \sim m_c$:

$$\hat{C}_1(\mu) = \frac{1}{2}\left[\left(\frac{\alpha_s(\mu)}{\alpha_s(m_b)}\right)^{-6/25}\left(\frac{\alpha_s(m_b)}{\alpha_s(M_W)}\right)^{-6/23} - \left(\frac{\alpha_s(\mu)}{\alpha_s(m_b)}\right)^{12/25}\left(\frac{\alpha_s(m_b)}{\alpha_s(M_W)}\right)^{12/23}\right], \tag{5.32}$$

$$\hat{C}_2(\mu) = \frac{1}{2}\left[\left(\frac{\alpha_s(\mu)}{\alpha_s(m_b)}\right)^{-6/25}\left(\frac{\alpha_s(m_b)}{\alpha_s(M_W)}\right)^{-6/23} + \left(\frac{\alpha_s(\mu)}{\alpha_s(m_b)}\right)^{12/25}\left(\frac{\alpha_s(m_b)}{\alpha_s(M_W)}\right)^{12/23}\right]. \tag{5.33}$$

Inserting numbers we find the values shown in Table 5.2.

For all other decay modes, we have a quark–antiquark pair of the same flavour involved; in particular, for all non-leptonic kaon decays, we have the quark transition $s \to u\bar{u}d$. These operators differ in their flavour structure: while in the operators previously analysed four quark flavours (two up-type flavours and two down-type flavours) change by one unit, in this case only two flavours (either two up-type or two down-type flavours) change by one unit. For these transitions, we have more dimension-six operators mixing with the initial operator. We shall discuss this issue by studying the quark transition $b \to s$, which involves two four-quark operators $b \to u\bar{u}s$ and $b \to c\bar{c}s$.

In this case the c quarks and the u quarks can form a loop and, together with the emission of a gluon, we can have the diagrams depicted in Fig. 5.2. The contribution to the effective Hamiltonian consists of the four-quark operators discussed above, which are

$$Q_1 = (\bar{c}_{\mathrm{L},i}\gamma_\mu b_{\mathrm{L},j})\,(\bar{s}_{\mathrm{L},j}\gamma_\mu c_{\mathrm{L},i}) \ , \tag{5.34}$$
$$Q_2 = (\bar{c}_{\mathrm{L},i}\gamma_\mu b_{\mathrm{L},i})\,(\bar{s}_{\mathrm{L},j}\gamma_\mu c_{\mathrm{L},j}) \ , \tag{5.35}$$
$$P_1 = (\bar{u}_{\mathrm{L},i}\gamma_\mu b_{\mathrm{L},j})\,(\bar{s}_{\mathrm{L},j}\gamma_\mu u_{\mathrm{L},i}) \ , \tag{5.36}$$
$$P_2 = (\bar{u}_{\mathrm{L},i}\gamma_\mu b_{\mathrm{L},i})\,(\bar{s}_{\mathrm{L},j}\gamma_\mu u_{\mathrm{L},j}) \tag{5.37}$$

for the c and u quarks; there will be further operators which the original operators given in (5.10) and (5.13) can mix into. These operators are usually called the QCD penguin operators [3, 4], and are given by

$$O_3 = (\bar{s}_{\mathrm{L},i}\gamma_\mu b_{\mathrm{L},i}) \sum_{q=u,d,s,c,b} (\bar{q}_{\mathrm{L},j}\gamma^\mu q_{\mathrm{L},j}) \ , \tag{5.38}$$

$$O_4 = (\bar{s}_{\mathrm{L},i}\gamma_\mu b_{\mathrm{L},j}) \sum_{q=u,d,s,c,b} (\bar{q}_{\mathrm{L},j}\gamma^\mu q_{\mathrm{L},i}) \ , \tag{5.39}$$

$$O_5 = (\bar{s}_{\mathrm{L},i}\gamma_\mu b_{\mathrm{L},i}) \sum_{q=u,d,s,c,b} (\bar{q}_{\mathrm{R},j}\gamma^\mu q_{\mathrm{R},j}) \ , \tag{5.40}$$

$$O_6 = (\bar{s}_{\mathrm{L},i}\gamma_\mu b_{\mathrm{L},j}) \sum_{q=u,d,s,c,b} (\bar{q}_{\mathrm{R},j}\gamma^\mu q_{\mathrm{R},i}) \ . \tag{5.41}$$

This contribution to the effective Hamiltonian for $b \to s = b \to s(\bar{u}u + \bar{c}c)$ takes the form

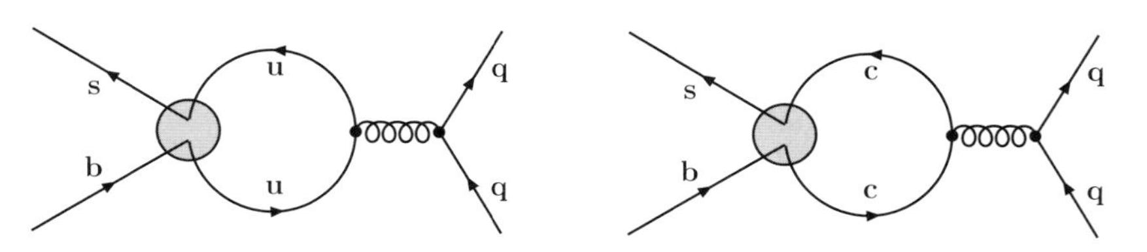

Fig. 5.2. QCD penguin diagrams for the transiton $b \to s$. Here q is any of the active quarks $q = u, d, s, c, b$

$$
\begin{aligned}
\mathcal{H}_{eff}(b \to s) = \frac{4G_F}{\sqrt{2}} \Big\{ & V_{cb}V_{cd}^* \left[C_1(\mu)Q_1(\mu) + C_2(\mu)Q_2(\mu)\right] \\
& + V_{ub}V_{ud}^* \left[C_1(\mu)P_1(\mu) + C_2(\mu)P_2(\mu)\right] \\
& - V_{tb}V_{td}^* \sum_{i=3}^{6} C_i(\mu)Q_i(\mu) \Big\} + \text{h.c.}
\end{aligned} \tag{5.42}
$$

The renormalization group equation allows us to calculate the Wilson coefficients of all the operators. Note that Q_1 and P_1 have the same coefficient in the same way as Q_2 and P_2, such that the anomalous dimension is now a 6×6 matrix, given at leading order by [1, 3, 4, 5, 6, 7, 8]

$$
\gamma = \frac{\alpha_s}{4\pi} \begin{pmatrix}
-2 & 6 & 0 & 0 & 0 & 0 \\
6 & -2 & -2/9 & 2/3 & -2/9 & 2/3 \\
0 & 0 & -22/9 & 22/3 & -4/9 & 4/3 \\
0 & 0 & 6-(2/9)n_f & -2+(2/3)n_f & -(2/9)n_f & (2/3)n_f \\
0 & 0 & 0 & 0 & 2 & -6 \\
0 & 0 & -(2/9)n_f & (2/3)n_f & -(2/9)n_f & -16+(2/3)n_f
\end{pmatrix}
+ \mathcal{O}(\alpha_s^2) \tag{5.43}
$$

where the upper left corner of this matrix has been considered already and yields the Wilson coefficients of the operators Q_1, P_1 and Q_2, P_2. Here, n_f denotes the number of "active" flavours.

The solution for the coefficients $C_i(\mu)$ cannot be given in a simple analytical form any more. When the coefficients are evolved from M_W down to m_b the numerical values of the coefficients given in Table 5.1 are obtained.

Table 5.1. Values of the Wilson coefficients $C_i(\mu)$ at three different scales of the order of the b quark mass, evaluated with the one-loop β function and the leading-order anomalous-dimension matrix (5.43), taking $\Lambda_{QCD} = 225$ MeV

$C_i(\mu)$	$\mu = 10.0\,\text{GeV}$	$\mu = 5.0\,\text{GeV}$	$\mu = 2.5\,\text{GeV}$
C_1	0.182	0.275	0.40
C_2	−1.074	−1.121	−1.193
C_3	−0.008	−0.013	−0.019
C_4	0.019	0.028	0.040
C_5	−0.006	−0.008	−0.011
C_6	0.022	0.035	0.055

All other flavour combinations (which are the $b \to c\bar{c}d$, $b \to u\bar{u}s$ and $b \to u\bar{u}d$ transitions) have the same Wilson coefficients and differ from the example discussed above only in the CKM factors and the flavours entering the four-quark operators.

For the non-leptonic interactions of charmed hadrons, one has to replace the quark flavours accordingly. By the same arguments, the transition $c \to u$ mediated by $c \to s\bar{s}u$ and $c \to d\bar{d}u$ will have penguin contributions. However, unlike in the case of a down-type quark, where the penguin contributions are driven by the large top-quark mass, we have for the charm-quark decays a perfect GIM cancellation (see Sect. 1.1) for all scales above the bottom-quark mass, since for such large scales all down-type quarks are treated as massless and the coefficients of the penguin operators vanish. A penguin contribution is induced at the scale of the b quark: when the b quark is integrated out, the GIM cancellation is incomplete and a penguin contribution is induced from the mixing of O_1 and O_2 into the penguin operators (5.38), where the appropriate replacements of the quark fields are understood. However, the 2×2 Cabbibo submatrix of the CKM matrix is orthogonal to a very good approximation, which makes the coefficients of the QCD penguins in charm decays very small; they are of the order

$$\frac{\alpha_s(m_b)}{\pi} V_{ub} V_{cb}^* \ln\left(\frac{m_b}{m_c}\right) ,$$

and thus we can safely ignore their contribution. Hence only the two Wilson coefficients C_1 and C_2 are relevant, the values of which are given for $\mu \sim m_c$ in Table 5.2.

Table 5.2. Values of the Wilson coefficients $C_i(\mu)$ at three different scales of the order of the c quark mass, evaluated with the one-loop β function and the leading-order anomalous-dimension matrix (5.43), taking $\Lambda_{QCD} = 215$ MeV

$C_i(\mu)$	$\mu = 1.0\,\mathrm{GeV}$	$\mu = 1.5\,\mathrm{GeV}$	$\mu = 2.0\,\mathrm{GeV}$
C_1	-0.60	-0.48	-0.41
C_2	1.32	1.24	1.20

In order to construct the effective hamiltonian for non-leptonic decays of kaons, one has to eventually integrate out also the charm quark. The effective Hamiltonian finally contains only the three light-quark flavours, and the relevant transition is $s \to u\bar{u}d$, which has penguin contributions enhanced by the large top mass. The renormalization group (with the same anomalous-dimension matrix, but with a reduced number of active flavours) can be used to evolve further down to small scales, but the (perturbative) renormalization group evolution has to be stopped at scales $\mu \sim 1\,\mathrm{GeV}$, since one enters the non-perturbative regime here. The relevant effective Hamiltonian can be written as

$$\mathcal{H}_{eff}(s \to d) = \frac{4G_F}{\sqrt{2}} V_{us} V_{ud}^* \sum_{i=1}^{6} \left[z_i(\mu) - \frac{V_{ts}}{V_{td}} y_i(\mu) \right] R_i(\mu) + \text{h.c.} \quad (5.44)$$

with the operators

$$R_1 = (\bar{u}_{\mathrm{L},i}\gamma_\mu s_{\mathrm{L},j}) (\bar{d}_{\mathrm{L},j}\gamma_\mu u_{\mathrm{L},i}) \ , \quad (5.45)$$

$$R_2 = (\bar{u}_{\mathrm{L},i}\gamma_\mu s_{\mathrm{L},i}) (\bar{d}_{\mathrm{L},j}\gamma_\mu u_{\mathrm{L},j}) \ , \quad (5.46)$$

$$R_3 = (\bar{d}_{\mathrm{L},i}\gamma_\mu s_{\mathrm{L},i}) \sum_{q=u,d,s} (\bar{q}_{\mathrm{L},j}\gamma^\mu q_{\mathrm{L},j}) \ , \quad (5.47)$$

$$R_4 = (\bar{d}_{\mathrm{L},i}\gamma_\mu s_{\mathrm{L},j}) \sum_{q=u,d,s} (\bar{q}_{\mathrm{L},j}\gamma^\mu q_{\mathrm{L},i}) \ , \quad (5.48)$$

$$R_5 = (\bar{d}_{\mathrm{L},i}\gamma_\mu s_{\mathrm{L},i}) \sum_{q=u,d,s} (\bar{q}_{\mathrm{R},j}\gamma^\mu q_{\mathrm{R},j}) \ , \quad (5.49)$$

$$R_6 = (\bar{d}_{\mathrm{L},i}\gamma_\mu s_{\mathrm{L},j}) \sum_{q=u,d,s} (\bar{q}_{\mathrm{R},j}\gamma^\mu q_{\mathrm{R},i}) \ . \quad (5.50)$$

The values of the Wilson coefficients z_i and y_i are tabulated in Table 5.3.

Table 5.3. Leading-log values of the Wilson coefficients $z_i(\mu)$ and $y_i(\mu)$ at three different scales around 1.5 GeV for $\Lambda_{QCD} = 215$ MeV. The entries marked with an asterisk are numerically irrelevant

	$\mu = 1.0\,\mathrm{GeV}$	$\mu = 1.3\,\mathrm{GeV}$	$\mu = 2.0\,\mathrm{GeV}$
z_1	−0.602	−0.518	−0.411
z_2	1.323	1.266	1.199
z_3	0.003	*	*
z_4	−0.008	*	*
z_5	0.003	*	*
z_6	−0.009	*	*
y_3	0.029	0.026	0.019
y_4	−0.051	−0.050	−0.040
y_5	0.012	0.013	0.011
y_6	−0.084	−0.075	−0.055

5.1.3 Electroweak Penguins

Finally we can also have so-called electroweak penguin operators, which are like the QCD penguins shown in Fig. 5.2, but with the gluon replaced by an electroweak boson, i.e. a photon or a Z_0. Although at first sight these contributions seem to be small owing to the smaller electroweak coupling, they can have a sizeable effect for down-type quarks owing to the large top-quark mass [9]. The structure of these operators is similar to that of the QCD penguins, but the couplings differ. For the decay of a bottom quark, we obtain

$$H_{\text{eff}}^{\text{EWP}}(b \to s) = \frac{4G_F}{\sqrt{2}} \alpha V_{ts} V_{tb}^* \sum_{i=7}^{10} C_i(\mu) O_i(\mu) , \tag{5.51}$$

with the operators

$$O_7 = \frac{3}{2} (\bar{s}_{\mathrm{L},i} \gamma_\mu b_{\mathrm{L},i}) \sum_{q=u,d,s,c,b} e_q (\bar{q}_{\mathrm{L},j} \gamma^\mu q_{\mathrm{L},j}) , \tag{5.52}$$

$$O_8 = \frac{3}{2} (\bar{s}_{\mathrm{L},i} \gamma_\mu b_{\mathrm{L},j}) \sum_{q=u,d,s,c,b} e_q (\bar{q}_{\mathrm{L},j} \gamma^\mu q_{\mathrm{L},i}) , \tag{5.53}$$

$$O_9 = \frac{3}{2} (\bar{s}_{\mathrm{L},i} \gamma_\mu b_{\mathrm{L},i}) \sum_{q=u,d,s,c,b} e_q (\bar{q}_{\mathrm{R},j} \gamma^\mu q_{\mathrm{R},j}) , \tag{5.54}$$

$$O_{10} = \frac{3}{2} (\bar{s}_{\mathrm{L},i} \gamma_\mu b_{\mathrm{L},j}) \sum_{q=u,d,s,c,b} e_q (\bar{q}_{\mathrm{R},j} \gamma^\mu q_{\mathrm{R},i}) , \tag{5.55}$$

where e_q is the charge of the quark q in units of the electron charge. Analogously, we can write the contributions for $b \to d$ transitions by making the replacement $s \to d$, but these amplitudes are CKM suppressed relative to $b \to s$ by a factor V_{td}/V_{ts}. The Wilson coefficients have been calculated to the next-to-leading logarithms and their values can be found in [2]. In Table 5.4, we give their values to leading logarithmic accuracy.

Table 5.4. Leading-log values of the Wilson coefficients $C_i(\mu)$ at the scale $\mu = m_b = 4.4\,\text{GeV}$ for $\Lambda_{QCD} = 225$ MeV; α is the electromagnetic coupling. The values are taken from [2]

$C_i(\mu)$	$\mu = 1.0\,\text{GeV}$
C_7/α	0.045
C_8/α	0.048
C_9/α	−1.280
C_{10}/α	0.328

Likewise, the operators for the decay of a strange quark are obtained by the replacement $b \to s$ and $s \to d$ in (5.52) – (5.55), yielding $R_7, \ldots, R_{10}$, and the electroweak penguin contribution to the $s \to d$ transition becomes

$$\mathcal{H}_{\text{eff}}^{\text{EWP}}(s \to d) = -\frac{4G_F}{\sqrt{2}} \alpha V_{ts} V_{td}^* \sum_{i=7}^{10} y_i(\mu) R_i(\mu) + \text{h.c.} , \tag{5.56}$$

where the coefficients are given in Table 5.5.

Table 5.5. Leading-log values of the Wilson coefficients $y_i(\mu)$ of the electroweak penguin contributions to $s \to d$ transitions for three different values of μ, for $\Lambda_{QCD} = 215$ MeV

	$\mu = 1.0\,\text{GeV}$	$\mu = 1.3\,\text{GeV}$	$\mu = 2.0\,\text{GeV}$
y_7/α	0.027	0.030	0.031
y_8/α	0.114	0.092	0.068
y_9/α	−1.491	−1.428	−1.357
y_{10}/α	0.650	0.558	0.442

5.1.4 Radiative and (Semi)leptonic Flavour-Changing Neutral-Current Processes

Another important class of $\Delta F = \pm 1$ processes is flavour-changing neutral-current (FCNC) processes with photons and leptons [10, 11]. As has been discussed above, such processes cannot happen at tree level in the Standard Model, but they are allowed at one-loop level, going through two charged-current vertices. An example of loop diagrams leading effectively to flavour-changing neutral currents, the relevant diagrams for the process $b \to s\gamma$ are depicted in Fig. 5.3.

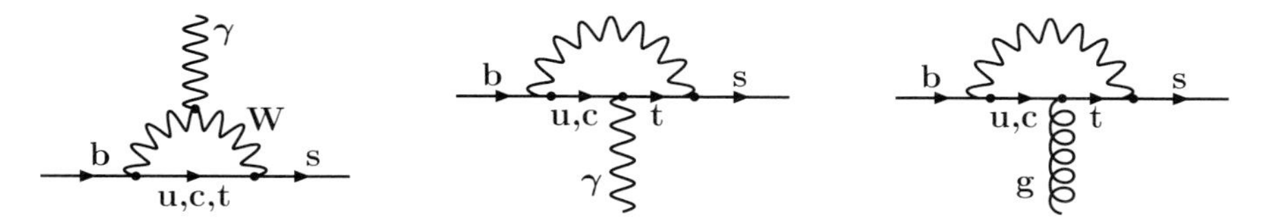

Fig. 5.3. Flavour-changing neutral-current loops for $b \to s\gamma$. Self-energy like diagrams are omitted. The third diagram contributes through its mixing

Although these loops diagram look divergent superficially, they actually lead to a convergent result once the summation over the flavours of the intermediate quark is performed. Considering again the transition $b \to s\gamma$, we can write the amplitude as

$$\mathcal{A}(b \to s\gamma) = V_{ub}V_{us}^* f(m_u) + V_{cb}V_{cs}^* f(m_c) + V_{tb}V_{ts}^* f(m_t)\,, \tag{5.57}$$

where $f(m)$ is the result of the loop integration, which depends on the quark mass m of the intermediate up-type quark. Any contribution which is the same for all intermediate quarks (i.e. a contribution independent of the quark mass) will cancel: if the quark masses of the up-type quarks were degenerate, $m_u = m_c = m_t = m$ the amplitude would vanish owing to CKM unitarity, i.e.

$$\mathcal{A}(b \to s\gamma) = f(m)\,[V_{ub}V_{us}^* + V_{cb}V_{cs}^* + V_{tb}V_{ts}^*] = 0\,, \tag{5.58}$$

which means that the amplitude is finite and proportional to the mass splitting of the up-type quarks, leading to an amplitude in our example roughly proportional to m_t^2. This is the essence of the GIM mechanism of the Standard Model discussed in Sect. 1.1, where FCNC processes are suppressed by loop factors $1/(16\pi^2)$.

Similar arguments hold for the FCNC transitions of s quarks, while for t or c quarks the intermediate quark is of down type. Since the mass splitting between the down-type quarks is much smaller, the GIM suppression for FCNC processes of up-type quarks is much more effective than for down-type quarks.

We can now proceed to apply the machinery of effective field theories to these processes. As above, we integrate out modes with masses of the order of those of the top quark and the weak boson. In total, we obtain the following for the contribution to the effective Hamiltonian:

$$H_{\text{eff}}(b \to s\gamma) = \frac{4G_F}{\sqrt{2}} V_{ts} V_{tb}^* \sum_{i=7,8} C_i'(\mu) P_i(\mu) , \qquad (5.59)$$

with the operators

$$P_7 = \frac{e}{16\pi^2} m_b (\bar{s}_{\text{L},\alpha} \sigma_{\mu\nu} b_{\text{R},\alpha}) F^{\mu\nu} , \qquad (5.60)$$

$$P_8 = \frac{g}{16\pi^2} m_b (\bar{s}_{\text{L},\alpha} T^a_{\alpha\beta} \sigma_{\mu\nu} b_{\text{R},\alpha}) G^{a\mu\nu} . \qquad (5.61)$$

Note that these contributions are proportional to the b quark mass. The corresponding Wilson coefficients at M_W are obtained by calculating the diagrams depicted in Fig. 5.3. Since the top quark appears in these diagrams, the results will be functions of the ratio m_t^2/M_W^2. These functions are called the Inami–Lim functions [12], and are, in the case at hand,

$$C_7'(M_W) = \frac{1}{2} x \left[\frac{2x^2/3 + 5x/12 - 7/12}{(x-1)^3} - \frac{3x^2/2 - x}{(x-1)^4} \ln x \right] , \qquad (5.62)$$

$$C_8'(M_W) = \frac{1}{2} x \left[\frac{x^2/4 - 5x/4 - 1/2}{(x-1)^3} + \frac{3x/2}{(x-1)^4} \ln x \right] , \qquad (5.63)$$

where $x = m_t^2/M_W^2$.

We can discuss FCNC quark-level processes involving leptons in the same way. We first consider the quark transition $b \to s\ell^+\ell^-$ where $\ell = e$, μ or τ. The relevant diagrams are shown in Fig. 5.4. Using the effective-field-theory picture and integrating out both the heavy top quark and the heavy weak bosons, we obtain two contributions

$$H_{\text{eff}}(b \to s\ell^+\ell^-) = \frac{4G_F}{\sqrt{2}} V_{ts} V_{tb}^* \sum_{i=9,10} C_i'(\mu) P_i(\mu) , \qquad (5.64)$$

with the operators

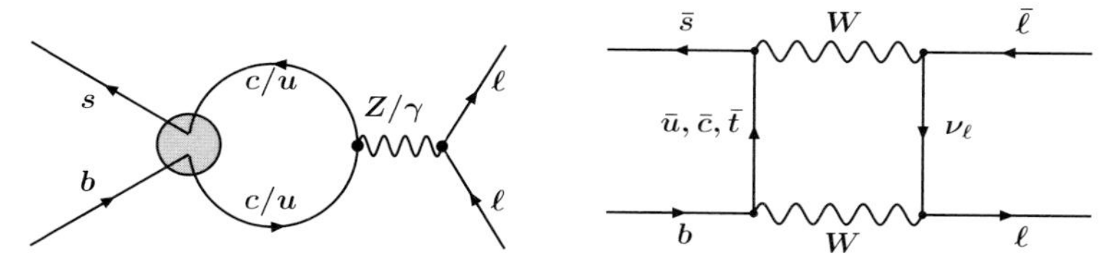

Fig. 5.4. Contributions to $b \to s\ell^+\ell^-$

$$P_9 = \frac{1}{2}(\bar{s}_{\rm L}\gamma_\mu b_{\rm L})(\bar{\ell}\gamma^\mu \ell) \,, \tag{5.65}$$

$$P_{10} = \frac{1}{2}(\bar{s}_{\rm L}\gamma_\mu b_{\rm L})(\bar{\ell}\gamma^\mu \gamma_5 \ell) \,. \tag{5.66}$$

The Wilson coefficients appearing in the effective Hamiltonian are obtained by calculating the diagrams in Fig. 5.4, from which we obtain another set of Inami–Lim functions,

$$\begin{aligned} C_9(M_W) &= \left(\frac{1}{\sin^2\theta_W}B(x) + \frac{-1+4\sin\theta_W}{\sin^2\theta_W}C(x) + D(x) + \frac{4}{9}\right) \\ C_{10}(M_W) &= \frac{-1}{\sin^2\theta_W}B(x) + \frac{1}{\sin^2\theta_W}C(x) \end{aligned} \tag{5.67}$$

where $x = m_t^2/M_W^2$, and we have defined the auxiliary functions

$$B(x) = \frac{1}{4}\left[\frac{-x}{x-1} + \frac{x}{(x-1)^2}\ln x\right] \,, \tag{5.68}$$

$$C(x) = \frac{x}{4}\left[\frac{x/2-3}{x-1} + \frac{3x/2+1}{(x-1)^2}\ln x\right] \,, \tag{5.69}$$

$$\begin{aligned} D(x) = &\left[\frac{-19x^3/36 + 25x^2/36}{(x-1)^3}\right. \\ &\left. + \frac{-x^4/6 + 5x^3/3 - 3x^2 + 16x/9 - 4/9}{(x-1)^4}\ln x\right] \,. \end{aligned} \tag{5.70}$$

As in the case of the other operators, we may obtain the expressions for the $b \to d\ell^+\ell^-$ and $s \to d\ell^+\ell^-$ transitions by the appropriate replacements.

The renormalization of these operators now involves also the operators O_i and P_i considered earlier. Diagrams such as those shown in Fig. 5.5 induce a mixing between the tree operators and the operators for $b \to s\gamma$ and $b \to s\ell^+\ell^-$. To perform this renormalization we have to introduce a 10×10 anomalous-dimension matrix which, is at order α_s, the extension of (5.43) and reads

$$\gamma = \frac{\alpha_s}{4\pi} \begin{bmatrix} -2 & 6 & 0 & 0 & 0 & 0 & 0 & 3 & -\frac{16}{3} & 0 \\ 6 & -2 & -\frac{2}{9} & \frac{2}{3} & -\frac{2}{9} & \frac{2}{3} & \frac{416}{81} & \frac{70}{27} & -\frac{16}{9} & 0 \\ 0 & 0 & -\frac{22}{9} & \frac{22}{3} & -\frac{4}{9} & \frac{4}{3} & -\frac{464}{81} & \frac{140}{27} + 3f & -\frac{16}{3}\left(u - \frac{d}{2} - \frac{1}{3}\right) & 0 \\ 0 & 0 & 6 - \frac{2}{9}f & -2 + \frac{2}{3}f & -\frac{2}{9}f & \frac{2}{3}f & \frac{416}{81}u - \frac{232}{81}d & 6 + \frac{70}{27}f & -\frac{16}{9}\left(u - \frac{d}{2} - 3\right) & 0 \\ 0 & 0 & 0 & 0 & 2 & -6 & \frac{32}{9} & -\frac{14}{3} - 3f & -\frac{16}{3}\left(u - \frac{d}{2}\right) & 0 \\ 0 & 0 & -\frac{2}{9}f & \frac{2}{3}f & -\frac{2}{9}f & -16 + \frac{2}{3}f & -\frac{448}{81}u + \frac{200}{81}d & -4 - \frac{119}{27}f & -\frac{16}{9}\left(u - \frac{d}{2}\right) & 0 \\ 0 & 0 & 0 & 0 & 0 & 0 & \frac{32}{3} & 0 & 0 & 0 \\ 0 & 0 & 0 & 0 & 0 & 0 & -\frac{32}{9} & \frac{28}{3} & 0 & 0 \\ 0 & 0 & 0 & 0 & 0 & 0 & 0 & 0 & -22 + \frac{4}{3}f & 0 \\ 0 & 0 & 0 & 0 & 0 & 0 & 0 & 0 & 0 & 0 \end{bmatrix} \tag{5.71}$$

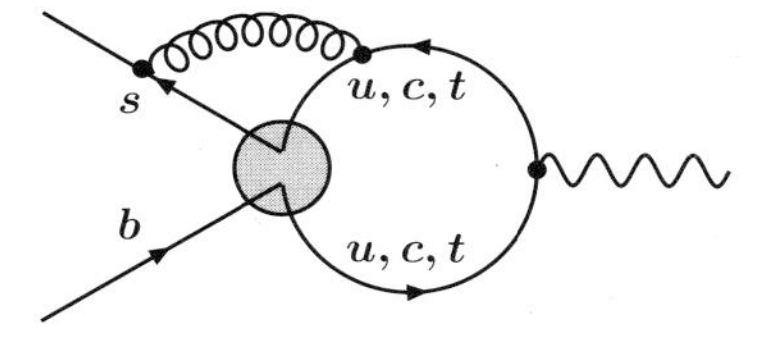

Fig. 5.5. Example of a diagramm leading to a mixing between O_2 and O_7

where f denotes the number of active flavours, and u and d are the number of active up and down flavours, respectively.

It should be noted that the mixing of the operators $O_1, \ldots, O_6$ into O_7 and O_8 depends on the regularization-scheme [13], and it is convenient to introduce, instead of the scheme-dependent coefficients C_7 and C_8, the scheme-independent ones

$$C_7^{\text{eff}}(\mu) = C_7(\mu) + \sum_{i=1}^{6} r_i C_i(\mu) , \tag{5.72}$$

$$C_8^{\text{eff}}(\mu) = C_8(\mu) + \sum_{i=1}^{6} s_i C_i(\mu) , \tag{5.73}$$

where the coefficients r_i and s_i depend on the regularization scheme. However, this is relevant only once next-leading-logarithms are included and we shall not discuss this any further; details can be found in [2].

By solving the renormalization group (4.15) with this anomalous-dimension matrix, we obtain the values for the Wilson coefficients shown in Table 5.6 [14].

Table 5.6. Values of the Wilson coefficients $C_i(\mu)$ for rare FCNC decays at three different scales of the order of the b quark mass. We have defined $C_9 = (\alpha/(2\pi))\hat{C}_9$. The values are taken from [2]

$C_i(\mu)$	$\mu = 2.5\,\text{GeV}$	$\mu = 5\,\text{GeV}$	$\mu = 10\,\text{GeV}$
C_7^{eff}	−0.334	−0.299	−0.268
C_8^{eff}	−0.157	−0.143	−0.131
$\hat{C}_9$	1.933	1.788	1.494

Note that the last column of the anomalous-dimension matrix (5.71) vanishes, which means that C_{10} is not renormalized. The numerical value is thus the same at all scales and is

$$C_{10}(\mu) = C_{10}(M_Z) = 4.69 \tag{5.74}$$

for a top-quark mass of 174 GeV.

Along the same, lines we can obtain the effective Hamiltonian for transitions of the form $s \to d\ell^+\ell^-$. The effective Hamiltonian reads

$$\mathcal{H}_{eff}^{\rm rare}(s \to d) = \frac{4G_F}{\sqrt{2}} V_{us}V_{ud}^*\alpha \sum_{i=9,10} \left[z_i'(\mu) - \frac{V_{ts}}{V_{td}} y_i'(\mu) \right] R_i'(\mu) + \text{h.c.} \,, \tag{5.75}$$

with the operators

$$R_9' = \frac{1}{2}(\bar{d}_{\rm L}\gamma_\mu s_{\rm L})(\bar{\ell}\gamma^\mu \ell) \,, \tag{5.76}$$

$$R_{10}' = \frac{1}{2}(\bar{d}_{\rm L}\gamma_\mu s_{\rm L})(\bar{\ell}\gamma^\mu \gamma_5 \ell) \,. \tag{5.77}$$

By the same argument as in the case of B decays the coefficient of R_{10}' is not renormalized; the other Wilson coefficients are tabulated in Table 5.7. We have $z_{10}' = 0$ and the value of y_{10}' is again not renormalized; $y_{10}' \approx 0.45$ for $m_t = 170\,\text{GeV}$.

Table 5.7. Leading-log values of the Wilson coefficients $z_9'(\mu)$ and y_9' for the rare FCNC processes mediated by $s \to d\ell^+\ell^-$ for three different values of μ and $\Lambda_{QCD} =$ 215 MeV

	$\mu = 0.8\,\text{GeV}$	$\mu = 1.0\,\text{GeV}$	$\mu = 1.2\,\text{GeV}$
z_9'/α	−0.031	−0.014	−0.004
y_9'/α	0.578	0.575	0.571

Finally, we may also discuss decays into a pair of neutrinos, e.g. the process $b \to s \sum \bar{\nu}\nu$, where the sum runs over the three neutrino species [15]. Here we may safely ignore the mass of the neutrinos and hence only left-handed neutrinos can appear. Consequently, we can have only a single operator

$$H_{\rm eff}(b \to s \sum \bar{\nu}\nu) = \frac{4G_F}{\sqrt{2}} V_{ts}V_{tb}^* \sum_\nu C'(\mu)P(\mu) \,, \tag{5.78}$$

with the operator

$$P = \frac{1}{2}(\bar{s}_L\gamma_\mu b_L)(\bar{\nu}\gamma^\mu(1-\gamma_5)\nu) \tag{5.79}$$

and the Wilson coefficient

$$C'(M_W) = \frac{\alpha}{2\pi \sin^2\theta_W} X\left(\frac{m_t^2}{M_W^2}\right) \tag{5.80}$$

where

$$X(x) = \frac{x}{8}\left[-\frac{2+x}{1-x} + \frac{3x-6}{(1-x)^2}\ln x \right] \,. \tag{5.81}$$

Note that the left-handed current $(\bar{s}_L \gamma_\mu b_L)$ is a conserved current in QCD, and so the operator P does not renormalize under QCD; thus

$$C'(\mu) = C'(M_W) = 0.008 \, . \tag{5.82}$$

The full effective Hamiltonian for all $\Delta F = 1$ processesis in fact known to next-to-leading-order accuracy. This has involved many contributions and also many technical details which are nicely reviewed in [2]. We shall return to some of the details in the sections where specific decay modes are discussed.

5.2 Remarks on $\Delta D = 1$ Processes: Pions and Nucleons

The hadrons consisting of up and down quarks are the lightest hadrons and hence no purely hadronic processes are possible. The leptonic and semileptonic processes of these hadrons were the first weak processes to have been observed, such as the decay of the charged pion and the β decay of the neutron. Clearly there is a very rich phenomenology of $d \to u$ transitions (for instance in nuclear β decays) which we cannot summarize here; we shall focus instead on a few very simple processes and present a few important examples.

Using the effective Hamiltonian constructed in the last section, we may start with the purely leptonic decays. The simplest process one may consider is the leptonic decay of the charged pion, $\pi^+ \to \mu^+ + \nu_\mu$ or $\pi^+ \to e^+ + \nu_e$. The necessary hadronic matrix element can be parametrized in terms of one constant f_π, the pion decay constant defined in (4.157) in Sect. 4.8.

The total decay rate can then be computed and yields

$$\Gamma(\pi^+ \to \mu^+ + \nu_\mu) = \frac{G_F^2}{4\pi} f_\pi^2 m_\mu^2 |V_{ud}|^2 \left(1 - \frac{m_\mu^2}{m_\pi^2}\right) \, . \tag{5.83}$$

For a given value of V_{ud}, this may be used to extract a value of f_π for which we obtain

$$f_\pi \approx 93\,\text{MeV} \, . \tag{5.84}$$

The interesting feature of (5.83) is its dependence on the mass of the lepton. Comparing $\pi^+ \to \mu^+ + \nu_\mu$ with $\pi^+ \to e^+ + \nu_e$ we find that the ratio of the two decay rates is

$$\frac{\Gamma(\pi^+ \to e^+ + \nu_e)}{\Gamma(\pi^+ \to \mu^+ + \nu_\mu)} = \frac{m_e^2}{m_\mu^2} \left(\frac{m_\pi^2 - m_e^2}{m_\pi^2 - m_\mu^2}\right)^2 \sim 1.283 \times 10^{-4} \, , \tag{5.85}$$

which is in accordance with the experimental findings.

This result is a consequence of the fact that the weak interaction is purely left-handed. Since the axial- and the vector currents of the leptons are conserved in the limit of vanishing lepton masses, the amplitude has to be proportional to the mass of the lepton. This result severely constrains possible scalar or tensor contributions to the weak hadronic current (see (1.2)).

A second process which can be calculated is the semileptonic decay of the charged pion $\pi^+ \to \pi^0 + e^+ + \nu_e$. In this case the hadronic matrix element is given in terms of two form factors $f_+(q^2)$ and $f_-(q^2)$

$$\langle \pi_0(p') | \bar{u}_L \gamma_\mu d_L | \pi^+(p) \rangle = f_+(q^2)(p_\mu + p'_\mu) + f_-(q^2) q_\mu \,, \tag{5.86}$$

where $q = p - p'$ is the momentum transferred to the leptons.

In order to deal with this unknown matrix element we make use of isospin symmetry which is a subgroup of the $SU(3)$ flavour symmetry discussed in Sect. 4.8. In the exact isospin limit the three pions would have identical masses and thus the decay $\pi^+ \to \pi^0 + e^+ + \nu_e$ would have no phase space. However, electromagnetic interactions break isospin by a small amount and thus the charged pion is slightly heavier than the neutral one. This mass difference is small, in which case we may neglect it in the form factors, which means that the form factors can be approximated by their value at $q^2 = 0$.

Writing the left-handed components in (5.86) explicitly we obtain two terms, one is the vector current and the other is the axial-vector current. Owing to parity, the axial current cannot contribute, and the vector current is one of the vector currents appearing in (4.156) which generates the (approximate) $SU(3)$ flavour symmetry. Thus this current is conserved, in which case the matrix element can be written in terms of a single form factor f_+:

$$\langle \pi_0(p') | \bar{u} \gamma_\mu d | \pi^+(p) \rangle \;. = f_+(q^2)(p_\mu + p'_\mu) \tag{5.87}$$

Furthermore, the corresponding charge is conserved, which fixes the value of f_+ at $q^2 = 0$:

$$f_+(q^2 = 0) = \sqrt{2} \,. \tag{5.88}$$

Using this value, we can compute the decay rate for the decay $\pi^+ \to \pi^0 + e^+ + \nu_e$ which becomes

$$\Gamma(\pi^+ \to \pi^0 + e^+ + \nu_e) = \frac{G_F^2 |V_{ud}|^2}{30\pi^3} (m_{\pi^+} - m_{\pi^0})^5 \,. \tag{5.89}$$

This result is indeed equal to the observed rate, supporting the hypothesis of the conserved vector current [16].

We can discuss the β decay of the neutron $n(p) \to p(p') e \bar{\nu}_e$ on the same footing. The transition matrix elements of the vector and axial-vector currents are parametrized in terms of six form factors in total:

$$\begin{aligned}
\langle n(p) | \bar{u} \gamma_\mu d | p(p') \rangle &= \bar{u}_n(p) \left[f_1(q^2) \gamma_\mu + f_2(q^2) \sigma_{\mu\nu} q^\nu + f_3 q_\mu \right] u_p(p') \,, \\
\langle n(p) | \bar{u} \gamma_\mu \gamma_5 d | p(p') \rangle & \\
&= \bar{u}_n(p) \left[g_1(q^2) \gamma_\mu + g_2(q^2) \sigma_{\mu\nu} q^\nu + g_3 q_\mu \right] \gamma_5 u_p(p') \,,
\end{aligned} \tag{5.90}$$

where u_p and u_n are the spinors for the proton and the neutron. As in the case of semileptonic π decay, the form factors are restricted by isospin or $SU(3)$ flavour symmetry. From the assumption that the vector current is conserved, we obtain $f_3(q^2) = 0$ in this limit. Furthermore, at vanishing momentum transfer, we obtain the following result from the fact that the vector current is conserved in the isospin limit (the CVC Hypothesis):

$$f_1(q^2 = 0) = 1 \ . \tag{5.91}$$

For the axial-vector form factors, we cannot use arguments from current conservation, since the axial vector is not conserved. The measured value [17]

$$G_A \equiv g_1(q^2 = 0) = 1.2715 \pm 0.0021 \tag{5.92}$$

can be explained using arguments from chiral perturbation theory. From these arguments one can obtain the Goldberger-Treiman relation for g_1 [18],

$$g_1(q^2 = 0) = \frac{g_{pn\pi} f_\pi}{m_p + m_n} \sim 1.31 \ , \tag{5.93}$$

where $g_{pn\pi}$ is the pion–nucleon coupling which is determined by experiment.

With this information, we can compute the rate for the β decay of the neutron. Again we make use of the argument that the form factors are needed only close to $q^2 = 0$; however, unlike in the case for the pions we may not neglect the electron mass here, since the mass difference between the neutron and the proton is of the order of the electron mass. A straightforward calculation yields

$$\begin{aligned} \frac{d^2\Gamma}{dE_e d\cos\theta} &= \frac{G_F^2 |V_{ud}|^2}{4\pi^3} (m_n - m_p - E_e)^2 E_e \sqrt{E_e^2 - m_e^2}\,\left[1 + 3G_A^2\right] \\ &\times \left[1 + \frac{\sqrt{E_e^2 - m_e^2}}{E_e} \frac{1 - G_A^2}{1 + 3G_A^2} \cos\theta \right] \ , \end{aligned} \tag{5.94}$$

where E_e is the electron energy and θ is the angle between the electron and the neutrino. Using this doubly differential rate allows a simultaneous extraction of both G_A and V_{ud}.

Finally, we mention some ways to determine the CKM matrix element V_{ud}. One precise way is to use super-allowed β decays of nuclei. Taking into account all known data, a recent article [19] quotes

$$|V_{ud}| = 0.9740 \pm 0.0005 \ . \tag{5.95}$$

Another way is to to use the β decay of the neutron discussed above [20]. However, in order to obtain a precise value, one has to put in the most precise value for the form factor $g_1(q^2 = 0)$ (5.92), and additional corrections such as the Coulomb distortion of the electron wave function when the electron is moving in the field of the proton. Taking into account all these effects yields

$$|V_{ud}| = 0.9728 \pm 0.0012 \ . \tag{5.96}$$

A summary of the current status can be found in the proceedings of a recent conference dedicated to the CKM matrix elements involving light quarks [21].

5.3 $\Delta S = 1$ Processes: Kaon Physics

Kaons and hyperons have a richer phenomenology, since the kaons are heavy enough to decay non-leptonically into two or three pions. We shall use the effective Hamiltonian to discuss some aspects of the phenomenology of kaon physics, starting from the simpler cases of leptonic and semileptonic decays, and then give a qualitative discussion of the non-leptonic channels. Unfortunately, the s-quark mass is not large enough to allow one to apply heavy mass methods, so one has either to resort to flavour $SU(3)$ and chiral perturbation theory (with the disadvantage that the breaking of $SU(3)$ is difficult to control) or to use models.

5.3.1 Leptonic and Semileptonic Kaon Decays

The simplest decay is the purely leptonic decay of the charged kaon which is computed in complete analogy to the decay of the charged pion. We find the following for the decay $K^+ \to \mu^+ + \nu_\mu$:

$$\Gamma(K^+ \to \mu^+ + \nu_\mu) = \frac{G_F^2}{4\pi} f_K^2 m_\mu^2 |V_{us}|^2 \left(1 - \frac{m_\mu^2}{m_K^2}\right) , \tag{5.97}$$

where the kaon decay constant is defined in a similar way to the pion decay constant:

$$\langle 0|\bar{u}_L \gamma_\mu s_L|K^+(p)\rangle = i f_K p_\mu . \tag{5.98}$$

Of course, this shows the same helicity suppression as in the case of pions, i.e. the rate is again proportional to the lepton mass.

If flavour $SU(3)$ (i.e. the symmetry between the s, u and d quarks) were an exact symmetry, we would have $f_\pi = f_K$. Clearly this symmetry is violated, since in this limit the pion mass would be equal to the kaon mass. However, the $SU(3)_{\text{flavour}}$ violation for the decay constants is at the level of 20%, which means that

$$f_K = (1.23 \pm 0.02) f_\pi \approx 114\,\text{MeV} . \tag{5.99}$$

We can discuss the semileptonic decay $K \to \pi \ell \bar{\nu}_\ell$, called the K_{e3} decay, in the same way. This decay is used to determine the CKM matrix element V_{us}. In order to compute the rate, we need the matrix element of the left-handed current between a kaon and a pion. Owing to parity, only the vector current can contribute; the matrix element of this current is parametrized in terms of two form factors F_+ and F_-:

$$\langle \pi_-(p')|\bar{u}\gamma_\mu s|K^0(p)\rangle = F_+(q^2)(p_\mu + p'_\mu) + F_+(q^2) q_\mu , \tag{5.100}$$

$$\langle \pi_0(p')|\bar{u}\gamma_\mu s|K^+(p)\rangle = \frac{1}{\sqrt{2}} F_+(q^2)(p_\mu + p'_\mu) + \frac{1}{\sqrt{2}} F_+(q^2) q_\mu , \tag{5.101}$$

where the factors of $\sqrt{2}$ emerge from applying isospin symmetry.

In contrast to the case of pions, the mass difference between the initial and the final state cannot be neglected. This has two consequences:

1. The q^2 dependence of the form factors cannot be neglected any more; hence we need some ansatz for the (non-perturbative) form factors.
2. The form factor F_- is non-zero and will contribute to the rate; however, its contribution in a semileptonic decay will be proportional to the lepton mass.

We shall neglect the lepton mass in the following, and hence only the first point needs to be considered.

The usual ansatz for the q^2 dependence of the form factor is based on the assumption that it is dominated by the propagator of the nearest resonance with the correct quantum numbers; this is commonly called the vector dominance model and is schematically depicted in Fig. 5.6. In the case of the $K \to \pi$ form factor this is the the K^* resonance, from which we obtain

$$F_+(q^2) = \frac{g_{KK^*\pi} f_{K^*}}{m_{K^*}^2 - q^2} , \tag{5.102}$$

which is called the pole ansatz for the form factor.

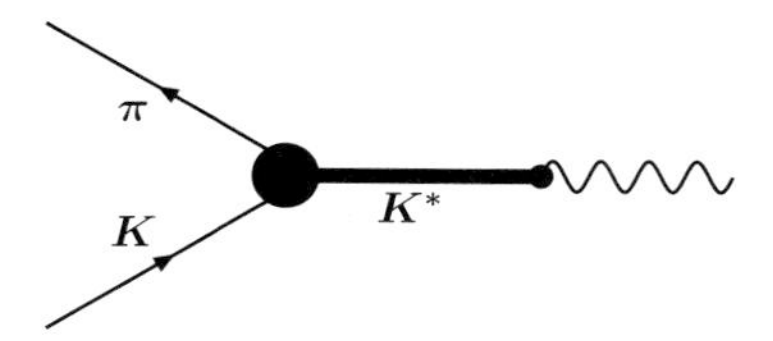

Fig. 5.6. Vector dominance model for the $K \to \pi$ form factor

Using this form factor, we can compute the rate for the decay $K \to \pi \ell \bar{\nu}_\ell$ in the limit of vanishing lepton mass. In order to be able to predict the (differential) rate quantitatively, we also need a value for $g_{KK^*\pi} f_{K^*}$, or, equivalently, for the form factor at $q^2 = 0$. This value can again be obtained from symmetry considerations, assuming exact $SU(3)_{\text{flavour}}$ symmetry, in which case we obtain $F_+(0) = 1$. However, this approximation is not as good as in the case of pions, since isospin breaking is much less than $SU(3)_{\text{flavour}}$ breaking. Thus, for a precise value for $F_+(0)$, one has to include corrections, which can be computed in the framework of chiral perturbation theory. This has been done in a classic paper by Leutwyler and Roos [22], who find

$$F_+(0) \approx 0.65 . \tag{5.103}$$

If we rely on the extrapolation given by the pole ansatz (5.102), we also obtain a prediction for the spectra (q^2 spectra as well as lepton energy spectra), which can be compared with experiment. The agreement is satisfactory and gives some confidence in the extraction of V_{us} from these decays.

In fact, the determination of V_{us} relies mainly on the measurement of these decays. The Particle Data Group [17] has extracted the value

$$V_{us} = 0.2196 \pm 0.0023 \tag{5.104}$$

which is consistent with other extractions, based on, for example, strange τ decays.

5.3.2 Non-Leptonic Kaon Decays

The kaon states are sufficiently heavy to decay into two pions and also into three pions, which is a non-leptonic decay. In terms of quarks, these transitions are mediated by the effective Hamiltonian considered in (5.44). However, in contrast to the leptonic and semileptonic decays, these transitions are much harder to describe, owing to hadronic uncertainties; in other words, the matrix elements of (5.44) are hard to calculate.

Before we consider the effective Hamiltonian, we start with a few more general considerations. As far as strong interactions are concerned, isospin is a good symmetry, and thus it is useful to discuss the various isospin components in the decay. The four kaon states K^0, K^+, $\overline{K}^0$ and K^- fall into two isodoublets, which are the CP conjugates of each other

$$\begin{pmatrix} K^+ \\ K^0 \end{pmatrix} \xrightarrow{\text{CP}} \begin{pmatrix} \overline{K}^0 \\ -K^- \end{pmatrix} . \tag{5.105}$$

For the neutral kaon states, we have[1]

$$\text{CP}|K^0\rangle = -|\overline{K}^0\rangle , \quad \text{CP}|\overline{K}^0\rangle = -|K^0\rangle . \tag{5.106}$$

Thus we can form CP eigenstates from the two neutral kaons, which are

$$|K_1\rangle = \frac{1}{\sqrt{2}} \left[|K^0\rangle - |\overline{K}^0\rangle \right] , \quad \text{CP}|K_1\rangle = |K_1\rangle , \tag{5.107}$$

$$|K_2\rangle = \frac{1}{\sqrt{2}} \left[|K^0\rangle + |\overline{K}^0\rangle \right] , \quad \text{CP}|K_2\rangle = -|K_2\rangle . \tag{5.108}$$

The final state can be either two or three pions. We focus first on the two-pion states; here the two pions have to be in a state with vanishing orbital angular momentum ℓ, since the kaon is a pseudoscalar particle. Such a state can be written as

$$|\pi^+\pi^-, \ell = 0\rangle = \int \frac{d\Omega_k}{\sqrt{4\pi}} \, |\pi^+(\boldsymbol{k})\pi^-(-\boldsymbol{k})\rangle , \tag{5.109}$$

[1] In fact, this implies a phase convention for the relative phase between the K^0 and $\overline{K}^0$ states. This convention is natural; the minus sign originates from the negative parity of the kaon states.

$$|\pi^0\pi^0, \ell = 0\rangle = \frac{1}{\sqrt{2}} \int \frac{d\Omega_k}{\sqrt{4\pi}} |\pi^0(\boldsymbol{k})\pi^0(-\boldsymbol{k})\rangle , \tag{5.110}$$

$$|\pi^+\pi^0, \ell = 0\rangle = \int \frac{d\Omega_k}{\sqrt{4\pi}} |\pi^+(\boldsymbol{k})\pi^0(-\boldsymbol{k})\rangle , \tag{5.111}$$

where the additional factor of $1/\sqrt{2}$ for the state with two neutral pions is due to the indistinguishability of the two particles.

We can now decompose the pion states into their various isopin components. A single pion has isospin $I = 1$ and consequently a two-pion state can have either $I = 0, 1$ or 2. Since the momentum state is symmetric for $\ell = 0$ (owing to the integration in (5.109)–(5.111), there is no way to couple the pions to give $I = 1$ owing to Bose symmetry: the total state has to be symmetric, and $I = 1$ corresponds to an antisymmetric combination of the two pions. Using the appropriate Clebsch–Gordan coefficients we find

$$\begin{aligned}
&|\pi(\boldsymbol{k})\pi(-\boldsymbol{k}), I = 0, I_3 = 0\rangle \\
&= \frac{1}{\sqrt{3}} \left[|\pi^+(\boldsymbol{k})\pi^-(-\boldsymbol{k})\rangle - |\pi^0(\boldsymbol{k})\pi^0(-\boldsymbol{k})\rangle + |\pi^-(\boldsymbol{k})\pi^+(-\boldsymbol{k})\rangle \right] , \quad (5.112) \\
&|\pi(\boldsymbol{k})\pi(-\boldsymbol{k}), I = 2, I_3 = 0\rangle \\
&= \frac{1}{\sqrt{6}} \left[|\pi^+(\boldsymbol{k})\pi^-(-\boldsymbol{k})\rangle + 2|\pi^0(\boldsymbol{k})\pi^0(-\boldsymbol{k})\rangle + |\pi^-(\boldsymbol{k})\pi^+(-\boldsymbol{k})\rangle \right] , \quad (5.113) \\
&|\pi(\boldsymbol{k})\pi(-\boldsymbol{k}), I = 2, I_3 = 1\rangle \\
&= \frac{1}{\sqrt{2}} \left[|\pi^+(\boldsymbol{k})\pi^0(-\boldsymbol{k})\rangle + |\pi^0(\boldsymbol{k})\pi^+(-\boldsymbol{k})\rangle \right] . \quad (5.114)
\end{aligned}$$

Inserting this into the states with orbital angular momentum $\ell = 0$,

$$|\pi\pi, I, I_3\rangle = \frac{1}{\sqrt{2}} \int \frac{d\Omega_k}{\sqrt{4\pi}} |\pi(\boldsymbol{k})\pi^0(-\boldsymbol{k})I, I_3\rangle , \tag{5.115}$$

we obtain

$$\begin{aligned}
|\pi^+\pi^-, \ell = 0\rangle &= \sqrt{\frac{2}{3}}|\pi\pi, I = 0, I_3 = 0\rangle + \sqrt{\frac{1}{3}}|\pi\pi, I = 2, I_3 = 0\rangle , \\
|\pi^0\pi^0, \ell = 0\rangle &= -\sqrt{\frac{1}{3}}|\pi\pi, I = 0, I_3 = 0\rangle + \sqrt{\frac{2}{3}}|\pi\pi, I = 2, I_3 = 0\rangle , \\
|\pi^+\pi^0, \ell = 0\rangle &= |\pi\pi, I = 2, I_3 = 1\rangle . \quad (5.116)
\end{aligned}$$

The effective Hamiltonian mediating a transition $K \to \pi\pi$ thus must have either $\Delta I = 1/2$, $\Delta I = 3/2$ or $\Delta I = 5/2$. The latter is not possible in the Standard Model, at least with a single insertion of the weak Hamiltonian discussed in Sect. 5.1. Omitting the possibility of a $\Delta I = 5/2$ contribution we can decompose the effective Hamiltonian into two pieces,

$$H_{eff} = H_{1/2} + H_{3/2} , \tag{5.117}$$

where both components have $I_3 = 1/2$. Using the Wigner–Eckart theorem, we can write the three decay amplitudes $A(K^0 \to \pi^+\pi^-)$, $A(K^0 \to \pi^0\pi^0)$ and $A(K^+ \to \pi^+\pi^0)$ in terms of two reduced matrix elements:

$$\begin{aligned}
\langle \pi\pi, I{=}0, I_3{=}0|H_{eff}|K^0\rangle &= \frac{1}{\sqrt{2}}\langle I{=}0||H_{1/2}||1/2\rangle \equiv A_0 \ , \\
\langle \pi\pi, I{=}2, I_3{=}0|H_{eff}|K^0\rangle &= -\frac{1}{\sqrt{10}}\langle I{=}2||H_{3/2}||1/2\rangle \equiv A_2 \ , \\
\langle \pi\pi, I{=}2, I_3{=}1|H_{eff}|K^+\rangle &= -\sqrt{\frac{3}{20}}\langle I{=}2||H_{3/2}||1/2\rangle = \sqrt{\frac{3}{2}}A_2 \ ,
\end{aligned} \quad (5.118)$$

where $\langle || \quad || \rangle$ denotes the reduced matrix elements appearing in the Wigner–Eckart theorem. We obtain the following for the decay amplitudes expressed in these reduced matrix elements

$$\begin{aligned}
\langle \pi^+\pi^0|H_{eff}|K^+\rangle &= +\sqrt{\frac{3}{2}}\,A_2 \ , \\
\langle \pi^0\pi^0|H_{eff}|K^0\rangle &= -\sqrt{\frac{1}{3}}\,A_0 + \sqrt{\frac{2}{3}}\,A_2 \ , \\
\langle \pi^+\pi^-|H_{eff}|K^0\rangle &= +\sqrt{\frac{2}{3}}\,A_0 + \sqrt{\frac{1}{3}}\,A_2 \ .
\end{aligned} \quad (5.119)$$

The absolute value of the ratio of the two reduced matrix elements can be inferred from the lifetimes and branching ratios of charged and neutral kaons. We shall neglect for the moment the small CP violation. Since all the two-pion states with $\ell = 0$ are CP even, i.e.

$$\mathrm{CP}|\pi\pi, \ell = 0\rangle = |\pi\pi, \ell = 0\rangle \ , \quad (5.120)$$

only the K_1 state given in (5.107) can decay into two pions, and the K_2 state has to decay into three pions. Since the phase space for the decay into three pions is much smaller than that for the decay into two pions, the lifetimes of the two states K_1 and K_2 are very different; if CP were conserved, we would have

$$|K_1\rangle = |K_{\mathrm{short}}\rangle \ , \quad |K_2\rangle = |K_{\mathrm{long}}\rangle \ , \quad (5.121)$$

where K_{short} is the short-lived and K_{long} the long-lived neutral kaon. However, CP is not conserved and we return to this matter in Chap. 6.

Neglecting the small violation of CP we can write

$$|K^0\rangle \approx \frac{1}{\sqrt{2}}\left[|K_{\mathrm{short}}\rangle + |K_{\mathrm{long}}\rangle\right] \ . \quad (5.122)$$

From this we obtain

$$\Gamma(K^+ \to \pi\pi) = \frac{1}{\tau(K^+)} \mathrm{Br}(K^+ \to \pi^+\pi^0) \propto \frac{3}{2}|A_2|^2 \,, \tag{5.123}$$

$$\begin{aligned}\Gamma(K^0 \to \pi\pi) &= \frac{1}{2\tau(K_{\text{short}})} \left[\mathrm{Br}(K_{\text{short}} \to \pi^0\pi^0) + \mathrm{Br}(K_{\text{short}} \to \pi^+\pi^-)\right] \\ &\propto |A_0|^2 + |A_2|^2 \,,\end{aligned} \tag{5.124}$$

where the factors of proportionality are the same up to phase space corrections, which are tiny. Inserting the numbers we can calculate the ratio

$$\frac{|A_0|}{|A_2|} = \sqrt{\frac{3}{2}\frac{\Gamma(K^0 \to \pi\pi)}{\Gamma(K^+ \to \pi\pi)} - 1} \;\approx\; 22 \,, \tag{5.125}$$

showing that the contribution of $H_{3/2}$ is strongly suppressed. This observation is one of the manifestations of the so-called $\Delta I = 1/2$ rule. This rule states that in decays of strange particles (Kaons and hyperons) the $\Delta I = 1/2$ contribution to the effective interaction is dominant in comparison with the $\Delta I = 3/2$ contribution; furthermore; no evidence for any $\Delta I = 5/2$ contribution has been found. As we shall see below, this substantial enhancement of the $\Delta I = 1/2$ interaction is one of the remaining mysteries; usually this is blamed on strong-interaction effects, which are hard to calculate.

As an interesting side remark, we may also perform a similar analysis on the D and B meson systems, since all the rates of the $D \to \pi\pi$ and $B \to \pi\pi$ decays have been measured. Using the lifetimes given in [17] and the recent measurement of the decay $B \to \pi^0\pi^0$, we obtain

$$\frac{|A_0^{D\to\pi\pi}|}{|A_2^{D\to\pi\pi}|} \approx 1.61 \,, \quad \frac{|A_0^{B\to\pi\pi}|}{|A_2^{B\to\pi\pi}|} \approx 0.92 \,, \tag{5.126}$$

and thus we observe no significant enhancement of the $\Delta I = 1/2$ contribution in the decays of heavier mesons. Hence the $\Delta I = 1/2$ enhancement must be considered to be accidental and may be related to the fact that the two-pion decays practically exhaust the inclusive non-leptonic rate, which is not the case for the decays of heavier hadrons.

On the basis of this isospin analysis, we can now examine the effective Hamiltonian discussed in Sect. 5.1 [23]. The tree-level Hamiltonian has the flavour structure (see (5.21))

$$H_{\text{eff}} \sim (\bar{u}s)(\bar{d}u) \tag{5.127}$$

where we have suppressed all Dirac and colour indices. Clearly this Hamiltonian has a $\Delta I = 1/2$ and a $\Delta I = 3/2$ contribution; however, this Hamiltonian does not exhibit a large enhancement of the $\Delta I = 1/2$ contribution over the $\Delta I = 3/2$ contribution.

It has been noted that once QCD corrections are switched on, the $\Delta S = \pm 1$ Hamiltonian receives contributions from penguins also, as has been discussed in Sect. 5.1. The contributions of the operators $R_3 \,, \ldots, \, R_6$

to (5.44) given in (5.47)–(5.50) involves a summation over the active quarks which are u, d and s in the present case. Thus the flavour structure (if we use the same "sloppy" notation as in (5.127)) of the penguin contributions (5.47)–(5.50) is

$$R_{3,\ldots,6} \sim (\bar{d}s)[(\bar{u}u) + (\bar{d}d) + (\bar{s}s)] \; . \tag{5.128}$$

The term in the square brackets is a flavour $SU(3)$ singlet and thus also an isosinglet, which adds the whole contribution from the QCD penguin operators to the $\Delta I = 1/2$ piece of the effective Hamiltonian. The contribution of the QCD penguins indeed leads to an enhancement of the $\Delta I = 1/2$ contribution compared with $\Delta I = 3/2$, but – depending on the scale used in the renormalization group evolution – the enhancement can be up to only a factor of about five; thus one cannot explain the observed factor of 22 by perturbative calculations and renormalization-group running [24].

5.4 $\Delta B = 1$ Processes: B Physics

The phenomenology of b hadrons is very rich; owing to the large mass of the bottom quark, there are many decay channels. On the other hand, the fact that the b quark mass is large compared with the scale parameter of QCD Λ_{QCD} allows us to perform a heavy-mass expansion. This method has been described in Sect. 4.3, and we shall discuss some of the main results in this section. The section is divided into subsections on exclusive and inclusive semileptonic decays, lifetimes, exclusive non-leptonic decays and rare decays.

5.4.1 Exclusive Semileptonic Decays

Semileptonic decays are mediated by the effective Hamiltonian shown in Sect. 5.1 and the remaining task is to consider the matrix elements of the hadronic currents of the $b \to c$ and $b \to u$ transitions between exclusive states.

We shall concentrate in this section on transitions of B mesons into the ground-state mesons D and D^*, mediated by the left-handed current $\bar{b}\gamma_\mu(1-\gamma_5)c$ These decays are the master example of a heavy-to-heavy transition, since we shall treat the c quark also as a heavy quark, i.e. we shall perform an expansion in $1/m_c$ also. This is in contrast to the transition $b \to u$, which is a heavy-to-light decay and in which heavy quark symmetries cannot be employed as efficiently.

The transition of a pseudoscalar into a pseudoscalar or vector meson is in general described in terms of six form factors, which we can write as

$$\langle D(v')|\bar{c}\gamma_\mu b|B(v)\rangle = \sqrt{m_B m_D}\left[\xi_+(y)(v_\mu + v'_\mu) + \xi_-(y)(v_\mu - v'_\mu)\right] \; , \tag{5.129}$$

$$\langle D^*(v',\epsilon)|\bar{c}\gamma_\mu b|B(v)\rangle = i\sqrt{m_B m_{D^*}}\,\xi_V(y)\varepsilon_{\mu\alpha\beta\rho}\epsilon^{*\alpha}v'^{\beta}v^{\rho} \; , \tag{5.130}$$

$$\begin{aligned}\langle D^*(v',\epsilon)|\bar{c}\gamma_\mu\gamma_5 b|B(v)\rangle = \sqrt{m_B m_{D^*}}\,&\left[\xi_{A1}(y)(vv'+1)\epsilon^*_\mu - \xi_{A2}(y)(\epsilon^* v)v_\mu\right.\\ &\left. -\xi_{A2}(y)(\epsilon^* v)v'_\mu\right] \; ,\end{aligned} \tag{5.131}$$

where we have defined $y = vv'$ and introduced convenient normalization factors. As we have discussed in Sect. 4.3, these six form factors are related to the Isgur–Wise function in the heavy-mass limit for both the b and the c quark. These relations are

$$\xi_i(y) = \xi(y) \quad \text{for } i = +, V, A1, A3\,, \qquad \xi_i(y) = 0 \quad \text{for } i = -, A2\,. \tag{5.132}$$

In particular, at the non-recoil point $v = v'$ we have due to heavy quark symmetry and Luke's theorem,

$$\begin{aligned} \xi_i(1) &= 1 + \mathcal{O}(1/m_Q^2) \quad \text{for } i = +, V, A1, A3\,, \\ \xi_i(1) &= \mathcal{O}(1/m_Q) \quad \text{for } i = -, A2\,. \end{aligned} \tag{5.133}$$

The differential rates for the exclusive semileptonic $b \to c$ transitions may be expressed in terms of the six form factors of (5.129)–(5.131) as

$$\begin{aligned} \frac{d\Gamma}{dy}(B \to D\ell\nu_\ell) &= \frac{G_F^2}{48\pi^3}|V_{cb}|^2(m_B + m_D)^2\left(m_D\sqrt{y^2-1}\right)^3 \\ &\quad \times \left|\xi_+(y) - \frac{m_B - m_D}{m_B + m_D}\xi_-(y)\right|^2 \end{aligned} \tag{5.134}$$

$$\begin{aligned} \frac{d\Gamma}{dy}(B \to D^*\ell\nu_\ell) &= \frac{G_F^2}{48\pi^3}|V_{cb}|^2(m_B - m_{D^*})^2 m_{D^*}^2\left(m_{D^*}\sqrt{y^2-1}\right) \\ &\quad \times (y+1)^2|\xi_{A1}(y)|^2 \sum_{i=0,\pm}|H_i(y)|^2 \end{aligned} \tag{5.135}$$

with the squared helicity amplitudes

$$|H_\pm(y)|^2 = \frac{m_B^2 - m_{D^*}^2 - 2ym_Bm_{D^*}}{(m_B - m_{D^*})^2}\left[1 \mp \sqrt{\frac{y-1}{y+1}}R_1(y)\right]^2\,, \tag{5.136}$$

$$|H_0(y)|^2 = \left(1 + \frac{m_B(y-1)}{m_B - m_{D^*}}\left[1 - R_2(y)\right]\right)^2\,. \tag{5.137}$$

Here we have defined the form factor ratios

$$R_1(y) = \frac{\xi_V(y)}{\xi_{A1}(y)}\,, \qquad R_2(y) = \frac{\xi_{A3}(y) + \frac{m_B}{m_{D^*}}\xi_{A2}(y)}{\xi_{A1}(y)}\,. \tag{5.138}$$

In the heavy-mass limit $m_b, m_c \to \infty$ these differential rates depend only on the Isgur–Wise function:

$$\frac{d\Gamma}{dy}(B \to D\ell\nu_\ell) \to \frac{G_F^2}{48\pi^3}|V_{cb}|^2(m_B + m_D)^2\left(m_D\sqrt{y^2-1}\right)^3|\xi(y)|^2\,, \tag{5.139}$$

$$\begin{aligned} \frac{d\Gamma}{dy}(B \to D^*\ell\nu_\ell) &\to \frac{G_F^2}{48\pi^3}|V_{cb}|^2(m_B - m_{D^*})^2 m_{D^*}^2\left(m_{D^*}\sqrt{y^2-1}\right)(y+1)^2 \\ &\quad \times \left[1 + \frac{4y}{y+1}\frac{m_B^2 - m_{D^*}^2 - 2ym_Bm_{D^*}}{(m_B - m_{D^*})^2}\right]|\xi(y)|^2\,. \end{aligned} \tag{5.140}$$

These relations allow a test of heavy-quark symmetry, since the ratios of the differential rates do not depend on any unknown form factor any more. In particular the ratios R_1 and R_2 measure the ratio of the differential transverse and longitudinal rates, respectively, to the total differential rate. In the heavy-mass limit both R_1 and R_2 are unity; this should be compared with the measurements by CLEO [25]

$$R_1 = 1.24 \pm 0.26 \pm 0.12 \, , \tag{5.141}$$

$$R_2 = 0.72 \pm 0.18 \pm 0.07 \, . \tag{5.142}$$

From the measured lepton invariant-mass spectrum, one may determine V_{cb} in a model independent way by extrapolating to the kinematical endpoint of maximal momentum transfer to the leptons, corresponding to the point $v = v'$. At this point heavy-quark symmetries determine the absolute normalization of some of the form factors, and the corrections to this normalization have been discussed in Sect. 4.3 and Sect. 4.4.

The mode $B \to D^* \ell \nu_\ell$ has the advantage of a higher branching fraction, and hence we shall start the discussion with this decay. The relevant formula may be derived from (5.135) and reads

$$\lim_{y \to 1} \frac{1}{\sqrt{y^2 - 1}} \frac{d\Gamma}{dy} (B \to D^* \ell \nu_\ell) = \frac{G_F^2}{4\pi^3} (m_B - m_{D^*})^2 m_{D^*}^3 |V_{cb}|^2 |\xi_{A1}(1)|^2 \, . \tag{5.143}$$

The form factor ξ_{A1} is normalized, owing to heavy-quark symmetries, and is hence protected against $1/m_Q$ corrections at $v = v'$ by Luke's theorem [26]. Hence we have

$$\xi_{A1}(1) = \eta_A (1 + \delta_{1/m^2}) \, . \tag{5.144}$$

Including QED corrections and the estimate of the $1/m_Q^2$ corrections in the way discussed in Sect. 4.3 and Sect. 4.4, we obtain [27]

$$\xi_{A1}(1) = 0.91^{+0.03}_{-0.04} \, . \tag{5.145}$$

For the extraction of V_{cb}, an extrapolation to the edge of phase space where $vv' = 1$ is necessary, and this extrapolation involves an assumption about the behaviour of the form factor $\xi_{A1}(vv')$ close to $vv' = 1$. The extrapolation usually uses a linear fit, in which the slope ρ^2 defined by

$$\xi_{A1}(vv') = \xi_{A1}(1) \left(1 - \rho^2 [vv' - 1] + \cdots \right) \tag{5.146}$$

is also extracted. From the theoretical side the slope is restricted by considerations of unitarity and analyticity [28, 29, 30]. The current results from the various experiments have been collected together and averaged by the Heavy Flavor Averaging Group [31] and are shown in Figs. 5.7 and 5.8. The value

$$V_{cb}^{\text{excl}} = (40.2 \pm 0.9_{\text{exp}} \pm 1.8_{\text{theo}}) \times 10^{-3} \tag{5.147}$$

has been extracted from these results [31].

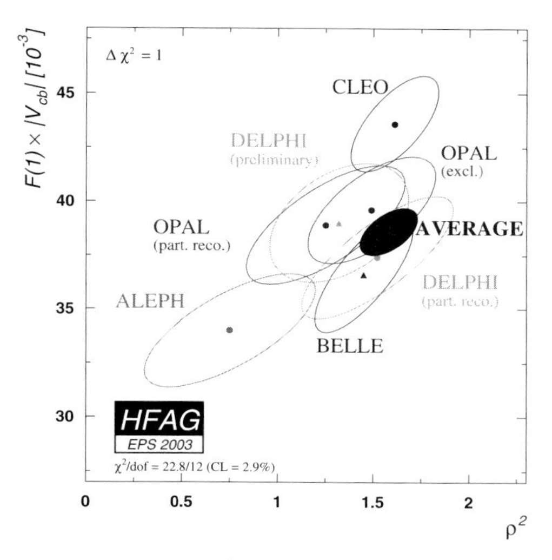

Fig. 5.7. Correlation ellipses in the ρ^2–$\xi_{A1}(1)V_{cb}$ plane. The plot is taken from [31]

The uncertainty quoted in (5.145) arises mainly from the unknown contributions of order $1/m_c^2$ and higher, and constitutes a limitation of this method until such time as lattice determinations improve our knowledge of the higher-order terms. Some initial progress has been made on this, see [32].

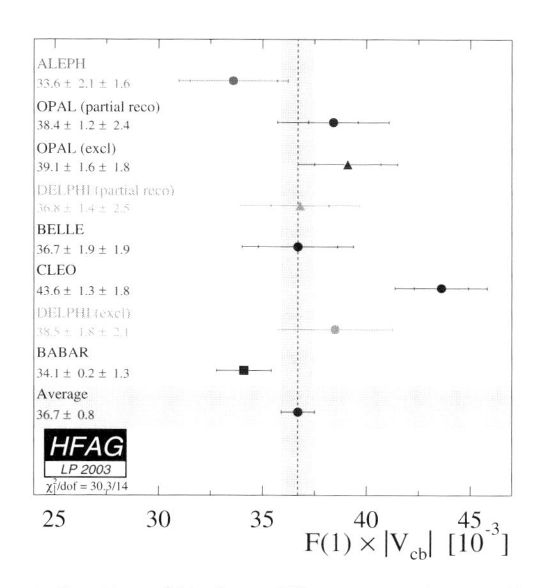

Fig. 5.8. Current situation of V_{cb} from different experiments. Results for $\xi_{A1}(1)V_{cb}$. The plot is taken from [31]

5.4.2 Inclusive Semileptonic Decays

Using the heavy-quark expansion described above, we may also calculate the rates for inclusive semileptonic decays. Starting from the simplest case, the total rate for $B \to X_c \ell \bar{\nu})\ell$, we may calculate straightforwardly the total rate at tree level, including the first, non-trivial non-perturbative corrections. We obtain the following for the total inclusive semileptonic decay rate $B \to X_c \ell \nu$:

$$\Gamma(B \to X_c \ell \nu)$$
$$\frac{G_F^2 m_b^5}{192\pi^3} |V_{cb}|^2 \left[\left(1 + \frac{\lambda_1}{2m_c^2}\right) f_1 \left(\frac{m_c}{m_b}\right) - \frac{9\lambda_2}{2m_c^2} f_2 \left(\frac{m_c}{m_b}\right) \right] , \quad (5.148)$$

where the two f_j are the phase-space functions

$$\begin{aligned} f_1(x) &= 1 - 8x^2 + 8x^6 - x^8 - 24x^4 \log x \, , \\ f_2(x) &= 1 - \frac{8}{3}x^2 - 8x^4 + 8x^6 + \frac{5}{3}x^8 + 8x^4 \log x \, . \end{aligned} \quad (5.149)$$

The result for $B \to X_u \ell \nu_\ell$ is obtained from (5.148) by taking the limit $m_c \to 0$ and making the replacement $V_{cb} \to V_{ub}$:

$$\Gamma(B \to X_u \ell \nu) = \frac{G_F^2 m_b^5}{192\pi^3} |V_{ub}|^2 \left[1 + \frac{\lambda_1 - 9\lambda_2}{2m_b^2} \right] . \quad (5.150)$$

As discussed above, the leading non-perturbative corrections in (5.148) and (5.150) are parametrized by λ_1 and λ_2. In order to estimate the total effect of the non-perturbative effects we shall insert a range of values $-0.3 > \lambda_1 > -0.6\,\mathrm{GeV}^2$; from this we obtain

$$\frac{\lambda_1 - 9\lambda_2}{2m_b^2} \sim -(3-4)\% \, . \quad (5.151)$$

This means that the non-perturbative contributions are small, in particular when compared with the perturbative ones, which were calculated some time ago [33, 34, 35, 36, 37, 38, 39]. For the decay $B \to X_u \ell \bar{\nu}_\ell$, the lowest order QCD corrections are given by

$$\begin{aligned} \Gamma(B \to X_u \ell \bar{\nu}_\ell) &= \frac{G_F^2 m_b^5}{192\pi^3} |V_{ub}|^2 \left[1 + \frac{2\alpha}{3\pi} \left(\frac{25}{4} - \pi^2 \right) \right] \\ &= 0.85 |V_{ub}|^2 \Gamma_b \, , \end{aligned} \quad (5.152)$$

and thus the typical size of QCD radiative corrections is of the order of 10–20%.

The method of the operator product expansion may also be used to obtain the non-perturbative corrections to the charged-lepton energy spectrum [40, 41, 42, 43, 44, 45]. In this case the operator product expansion is applied not to the full effective Hamiltonian, but only to the hadronic currents. The rate is written as a product of the hadronic tensor $W_{\mu\nu}$ and the leptonic tensor $\Lambda^{\mu\nu}$,

$$d\Gamma = \frac{G_F^2}{4m_B}|V_{Qq}|^2 W_{\mu\nu}\Lambda^{\mu\nu} d(PS) , \tag{5.153}$$

where $d(PS)$ is the phase-space differential. The short-distance expansion is then performed for the two currents appearing in the hadronic tensor. After the phase of the heavy-quark fields is redefined as in (4.46), it is found that the momentum transfer variable relevant for the short-distance expansion is $m_Q v - q$, where q is the momentum transfer to the leptons. After integration over the momentum of the neutrino, the expansion variable is the energy release $(m_Q/2) - E_\ell$, where E_ℓ is the energy of the charged lepton.

The structure of the expansion for the spectrum is identical to that for the total rate. The contribution of the dimension-three operators yields the free-quark decay spectrum, there are no contributions from dimension-four operators, and the $1/m_b^2$ corrections are parametrized in terms of λ_1 and λ_2. The result has already been shown in (4.91). It is interesting to note that even for finite charm mass the behaviour iat the endpoint is unphysical; this becomes manifest when we take the limit $m_c \to 0$ given in (4.94).

Figure 5.9 shows the distributions for inclusive semileptonic decays of B mesons. The spectrum close to the endpoint, where the lepton energy becomes maximal, exhibits a sharp spike as $y \to y_{max}$. In this region we have

$$\frac{d\Gamma}{dy} \propto \Theta(1-y-\rho) \left[2 + \frac{\lambda_1}{(m_Q(1-y))^2}\left(\frac{\rho}{1-\rho}\right)^2 \left\{3 - 4\left(\frac{\rho}{1-\rho}\right)\right\}\right] , \tag{5.154}$$

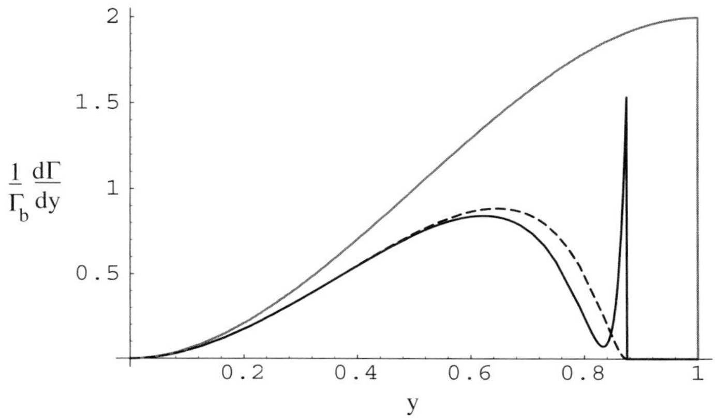

Fig. 5.9. The electron spectrum for free-quark $b \to c$ decay (*dashed line*), free-quark $b \to u$ decay (*grey line*) and $B \to X_c e\bar{\nu}_e$ decay including $1/m_b^2$ corrections (*solid line*) with $\lambda_1 = -0.5\,\mathrm{GeV}^2$ and $\lambda_2 = 0.12\,\mathrm{GeV}^2$. The figure is from [44]

which behaves like a δ-function and its derivative as $\rho \to 0$; this can be seen in (4.94). This behaviour indicates a breakdown of the operator product expansion close to the endpoint, since the expansion parameter for the spectra is not $1/m_Q$, but rather $1/(m_Q - qv)$, which becomes $1/(m_Q[1-y])$ after the integration over the neutrino momentum. In order to obtain a description of the endpoint region, one has to perform a resummation of the operator product expansion.

This resummation is the twist expansion discussed in Sect. 4.6. The spectrum may be calculated in terms of the distribution function f introduced in Sect. 4.6. It is interesting to note that, to leading order in the twist expansion, one can write the result as a convolution with an "effective mass" m_b^*; one finds (assuming the final-state quark to be massless) that [46]

$$\frac{d\Gamma}{dE_\ell} = \frac{G_F^2 \mid V_{qb}\mid^2}{12\pi^3} E_\ell^2 \int dk_+ \, f(k_+) \, \Theta\Big[m_b^* - 2\,E_\ell\Big]\Big\{3m_b^{*2} - 4m_b^* \, E_\ell\Big\} \,, \tag{5.155}$$

where $m_b^* = m_b + k_+$.

Note that the heavy quark mass m_b no longer appears explicitly. For this reason, in particular when the focus is on the endpoint region, it would be unnatural to introduce the rescaled lepton energy $y = 2E_\ell/m_b$. Hence, we shall hereafter present our results as a function of the lepton energy E_ℓ, which is the quantity that is actually measured in experiments. Note in particular, that (to the order we are working) the maximum value of the lepton energy is correctly reproduced.

The result (5.155) represents a resummation of the most singular contributions in the endpoint region, which corresponds to a resummation of the highest derivatives of δ-functions that appear in (4.94), where the explicit calculation to order $1/m_Q^2$ has been performed. In order to illustrate the effect of the convolution (5.155), we show in Fig. 5.10 the spectrum for $B \to X_u \ell \nu_\ell$ obtained using the ansatz [46]

$$f(k_+) = \frac{32}{\pi^2 \bar{\Lambda}} (1-x)^2 \exp\left\{ -\frac{4}{\pi} (1-x)^2 \right\} \Theta(1-x) \,, \tag{5.156}$$

where $x = k_+/\bar{\Lambda}$, and the choice $\bar{\Lambda} = 570$ MeV yields reasonable values for the moments.

Including the non-perturbative effects yields a reasonably behaved spectrum in the endpoint region and the δ-function-like singularities have disappeared. Furthermore, the spectrum now extends beyond the parton model endpoint; it is shifted from $E_\ell^{max} = m_Q/2$ to the physical endpoint $E_\ell^{max} = M_H/2$, since f is non-vanishing for positive values of $k_+ < \bar{\Lambda} = M_H - m_Q$.

Using the relations from Sect. 4.6, we may also include subleading light-cone distributions. In terms of the functions defined in (4.128), (4.127) and (4.122) we obtain the following for the lepton energy spectrum:

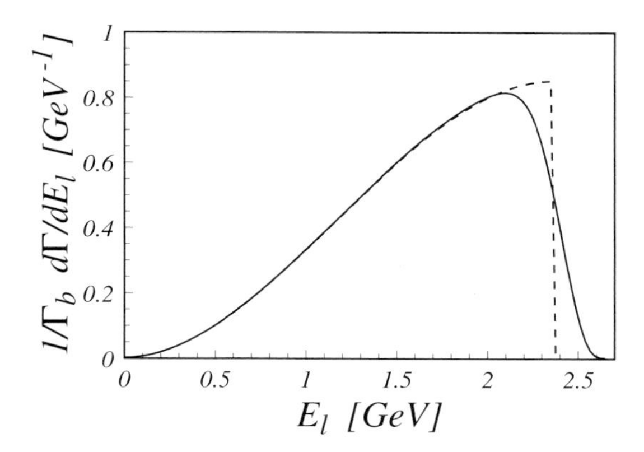

Fig. 5.10. Charged-lepton spectrum for $B \to X_u \ell \bar{\nu}$ decays. The *solid line* is (5.155) with the ansatz (5.156), and the *dashed line* shows the prediction of the free-quark decay model. The figure is from [46]

$$\frac{d\Gamma}{dE_\ell} = \frac{2\Gamma_0}{m_b} \int d\omega\, \theta(m_b - 2E_\ell - \omega)$$
$$\times \left[F(\omega) \left(1 - \frac{\omega}{m_b}\right) - \frac{1}{m_b} h_1(\omega) + \frac{3}{m_b} H_2(\omega) \right] + \mathcal{O}\left(\frac{\Lambda_{\mathrm{QCD}}^2}{m_b^2}\right) . \quad (5.157)$$

However, in order to exploit this relation for, for example, a model independent determination of V_{ub} one needs to know something about the subleading light-cone distributions. As has been discussed in [47], a knowledge of the first moments of the subleading functions may be sufficient to obtain a reasonably precise calculation of the rate close to the endpoint.

Currently, the most precise determination of V_{ub} relies on inclusive channels, since any exclusive determination still suffers from hadronic uncertainties that are larger than those in the inclusive calculation. In the inclusive case one is restricted by a possibly large charm background, which can be suppressed by various methods. One model-independent possibility is to compare the lepton energy spectrum of $B \to X_u \ell \bar{\nu}_\ell$ with the photon energy spectrum of $B \to X_s \gamma$, which is directly proportional to $f(\omega)$. Although this comparison can be performed, including even the subleading twist terms [47, 48, 49], it still needs the function $f(\omega)$ as an input, which increases the theoretical uncertainty of this method. Furthermore, only a small fraction, about 10% of the rate is actually above the cut at $E_\ell > (M_B^2 - M_D^2)/(2M_B)$ which is needed to get rid of the charm background, so this method certainly has serious drawbacks.

The advantage of the cut on the lepton energy spectrum is that the neutrino momentum does not need to be reconstructed. However, once the neutrino momentum is known, one may also use other variables to perform cuts. One alternative is the hadronic invariant mass m_X^2 [50, 51], which is peaked at small values for charmless decays and thus may serve as a very efficient cut. However, although in this case about 80% of the rate is still within a cut $M_X^2 < M_D$, there is still a dependence on the light-cone distribution function. Another alternative is to cut on the leptonic invariant mass q^2 [52, 53], which still has about 20% of the rate within the cut $q^2 > (M_B - M_D)^2$, however, this method does not depend on the light-cone distribution function. The effect of the cuts is shown schematically in Fig. 5.11. The shaded bar indicates the region which needs to be cut away in order to suppress the $b \to c$ background.

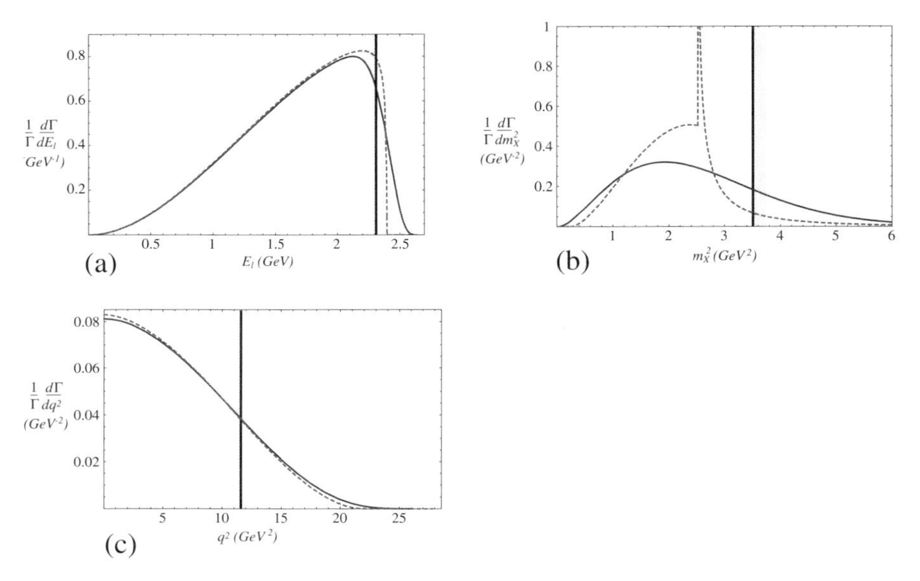

Fig. 5.11. Effect of the different possible cuts on $b \to u$ semileptonic decays. The region indicated by the *shaded bar* is contaminated by $b \to c$ decays. The *solid line* is the tree-level result (including the effect of the light-cone distribution function); the *dashed line* includes QCD radiative corrections. Plot (**a**) is the lepton energy spectrum, plot (**b**) is the hadronic invariant-mass spectrum and plot (**c**) is the leptonic invariant-mass spectrum. The figure is taken from [27]

By combining these different cuts in an optimized way one can arrive at a scheme which still includes about 45% of the $b \to u\ell\bar{\nu}_\ell$ rate and has only a moderate dependence on the light-cone distribution function [54]. In such a scheme, a theoretical uncertainty of $\Delta V_{ub}/V_{ub} \sim 5\%$ seems to be achievable at the B factories.

On top of the large non-perturbative corrections in the endpoint region, there are also large perturbative corrections, which have been known for a long time [33]. While the corrections to the total rate are of the expected size (see 5.152)), the endpoint region contains large logarithms, of which the leading ones are the Sudakov-like double logarithms. Up to terms vanishing at the endpoint, we have, to order α_s

$$\frac{d\Gamma}{dy} = \frac{d\Gamma^{(0)}}{dy}\left[1 - \frac{2\alpha_s}{3\pi}\left(\ln^2(1-y) + \frac{31}{6}\ln(1-y) + \frac{5}{4} + \pi^2\right)\right] , \tag{5.158}$$

where $y = 2E_\ell/m_b$ is again the rescaled energy variable and $\Gamma^{(0)}$ is the parton model rate which becomes close to the endpoint

$$\frac{1}{\Gamma_b}\frac{d\Gamma^{(0)}}{dy} = 2y^2(3-2y)\Theta(1-y) \to 2\Theta(1-y) . \tag{5.159}$$

The doubly logarithmic terms are believed to exponentiate, yielding

$$\frac{d\Gamma}{dy} = \frac{d\Gamma^{(0)}}{dy}\exp\left[-\frac{2\alpha_s}{3\pi}\ln^2(1-y)\right] , \tag{5.160}$$

up to singly logarithmic terms which have been neglected in (5.160). Note that the exponentiation strongly reduces the rate close to the endpoint, resulting in a strong modification of the parton model result.

Since singly logarithmic terms are omitted, the choice of the scale μ in α_s is not obvious, and various suggestions have been made. In particular, the choice $\mu^2 = m_b^2(1-y)$ yields a drastic modification of the spectrum in the endpoint region; using the one-loop result for α_s, we have

$$\frac{d\Gamma}{dy} = \frac{d\Gamma^{(0)}}{dy}\exp\left[-\frac{8}{25}\frac{\ln^2(1-y)}{\ln[(m_b^2/\Lambda_{QCD}^2)(1-y)]}\right] , \tag{5.161}$$

yielding a strong damping in the endpoint region, called the Sudakov suppression.

Unlike the non-perturbative corrections, the perturbative corrections cannot shift the endpoint of the spectrum away from the value given by the parton model. However, both types, the perturbative and the non-perturbative corrections, are strongly entangled in the endpoint region and a simultaneous treatment of both is difficult [55].

5.4.3 Lifetimes of $B^\pm$, B^0 and Λ_b

The subject of heavy-hadron lifetimes is strongly related to non-leptonic processes, which were considered some time ago [56, 57]; however, the systematic application of the $1/m_Q$ expansion has turned many assumptions into quantitative arguments. As outlined in the last section, the $1/m_Q$ expansion allows one to calculate total rates, even for non-leptonic processes, and hence a QCD-based calculation of lifetimes becomes possible.

Mainly as a result of the LEP experiments, the data on b-hadron lifetimes are quite precise. In particular, the lifetime ratios are known with precisions well below 10% [17]:

$$\frac{\tau(B^-)}{\tau(B_d)} = 1.062 \pm 0.029\,,$$
$$\frac{\tau(B_s)}{\tau(B_d)} = 0.964 \pm 0.045\,,$$
$$\frac{\tau(\Lambda_b)}{\tau(B_d)} = 0.780 \pm 0.037\,. \tag{5.162}$$

where $\tau(B_s)$ is the averaged B_s-meson lifetime.

To leading order in the $1/m_Q$ expansion, all lifetime ratios are unity, since the lifetimes are then determined by the free-quark decay. If we assume light-flavour symmetry for the matrix elements λ_1 and λ_2, the correction to the ratio between two meson lifetimes has to be of order $1/m_Q^3$ and hence is given by dimension-six operators. Since the value of λ_1 is different for a meson and a baryon and $\lambda_2 = 0$ for a Λ_Q baryon, we expect

$$\frac{\tau(B^-)}{\tau(B_d)} = 1 + O(1/m_b^3)\,,$$
$$\frac{\tau(B_s)}{\tau(B_d)} = 1 + O(1/m_b^3)\,,$$
$$\frac{\tau(\Lambda_b)}{\tau(B_d)} = 1 + O(1/m_b^2) \tag{5.163}$$

and we can recognize a potential problem with the lifetime ratio $\tau(\Lambda_b)/\tau(B_d)$. We shall return to this matter at the end of this subsection.

In order to discuss the lifetime ratios for mesons, one needs to consider higher orders in the $1/m_Q$ expansion. Unfortunately, in general the number of dimension-six operators that appear at order $1/m_Q^3$ is quite large. However, not all of these operators contribute to lifetime ratios, since they have to be sensitive to the light quark inside the heavy hadron.

These types of spectator effects have been reconsidered recently in [58], and we shall follow the line of argument given there. Although intuitively these dimension-six operators might seem to yield only small contributions, their effect is enhanced by a phase space factor. The non-spectator effects typically yield, at the partonic level, at least a three-particle phase space, while the spectator effects have, to leading order only, a two particle final state. This yields an enhancement of these contributions by a phase space factor $16\pi^2$. Qualitatively, this means that the $1/m_Q$ expansion for the lifetime ratios takes the form

$$\frac{\tau(B^-)}{\tau(B_d)} = 1 + \frac{1}{m_b^3}\left[a_0 + \frac{1}{m_b}a_1 + \cdots\right] + \frac{16\pi^2}{m_b^3}\left[b_0 + \frac{1}{m_b}b_1 + \cdots\right]\,, \tag{5.164}$$

and hence the coefficient b_0 is the leading contribution to the lifetime ratio.

We start from the effective Hamiltonian for non-leptonic weak decays given in Sect. 5.1. Performing the steps to construct a $1/m$ expansion for the total non-leptonic rate, we obtain first the partonic result, i.e. the lifetime for the free-quark decay, which is the first term in (5.164). The first non-vanishing terms will be the same for the neutral and the charged B meson, and hence the first terms will arise from dimension-six operators. The correction to the decay rate due to the spectator effects may be expressed in terms of four-quark operators, which we choose to be

$$\begin{aligned} O^q_{V-A} &= \bar{b}_L \gamma_\mu q_L \, \bar{q}_L \gamma^\mu b_L \,, \\ O^q_{S-P} &= \bar{b}_R \, q_L \, \bar{q}_L \, b_R \,, \\ T^q_{V-A} &= \bar{b}_L \gamma_\mu t_a q_L \, \bar{q}_L \gamma^\mu t_a b_L \,, \\ T^q_{S-P} &= \bar{b}_R \, t_a q_L \, \bar{q}_L \, t_a b_R \,, \end{aligned} \tag{5.165}$$

where $t_a = \lambda_a/2$ are the generators of colour $SU(3)$. In terms of these operators, the relevant contribution to the total decay rate becomes [58]

$$\begin{aligned} \Gamma = \Gamma_0 &+ \frac{2G_F^2 m_b^2}{\pi} |V_{cb}|^2 (1-z)^2 \\ &\left\{ \left(2c_1c_2 + \frac{1}{N_c}(c_1^2 + c_2^2) \right) \langle O^u_{V-A} \rangle + 2(c_1^2 + c_2^2) \langle T^u_{V-A} \rangle \right\} \\ &- \frac{2G_F^2 m_b^2}{3\pi} |V_{cb}|^2 (1-z)^2 \left\{ \left(2c_1c_2 + \frac{1}{N_c} c_1^2 + N_c c_2^2 \right) \right. \\ &\left[\left(1 + \frac{z}{2}\right) \langle O^{d'}_{V-A} \rangle - (1+2z) \langle O^{d'}_{S-P} \rangle \right] \\ &\qquad \left. + 2c_1^2 \left[\left(1 + \frac{z}{2}\right) \langle T^{d'}_{V-A} \rangle - (1+2z) \langle T^{d'}_{S-P} \rangle \right] \right\} \\ &- \frac{2G_F^2 m_b^2}{3\pi} |V_{cb}|^2 \sqrt{1-4z} \\ &\left\{ \left(2c_1c_2 + \frac{1}{N_c} c_1^2 + N_c c_2^2 \right) \left[(1-z) \langle O^{s'}_{V-A} \rangle - (1+2z) \langle O^{s'}_{S-P} \rangle \right] \right. \\ &\qquad \left. + 2c_1^2 \left[(1-z) \langle T^{s'}_{V-A} \rangle - (1+2z) \langle T^{s'}_{S-P} \rangle \right] \right\} , \end{aligned} \tag{5.166}$$

where $z = m_c^2/m_b^2$, and $N_c = 3$ is the number of colours.

We shall focus here on the mesonic case; the baryonic case is discussed in [58]. The matrix elements of the four-quark operators are parametrized as

$$\frac{1}{2m_{B_q}} \langle B_q | O^q_{V-A} | B_q \rangle \equiv \frac{f^2_{B_q} m_{B_q}}{8} B_1 \,,$$
$$\frac{1}{2m_{B_q}} \langle B_q | O^q_{S-P} | B_q \rangle \equiv \frac{f^2_{B_q} m_{B_q}}{8} B_2 \,,$$
$$\frac{1}{2m_{B_q}} \langle B_q | T^q_{V-A} | B_q \rangle \equiv \frac{f^2_{B_q} m_{B_q}}{8} \varepsilon_1 \,,$$
$$\frac{1}{2m_{B_q}} \langle B_q | T^q_{S-P} | B_q \rangle \equiv \frac{f^2_{B_q} m_{B_q}}{8} \varepsilon_2 \,, \tag{5.167}$$

where this parametrization is motivated by naive factorization, which will be discussed in connection with exclusive non-leptonic decays at the end of this section. The assumption of factorization yields $B_i = 1$ and $\varepsilon_i = 0$. Naive factorization holds exactly in the limit of large N_c and hence we expect

$$B_i = O(1) \,, \qquad \varepsilon_i = O(1/N_c) \,. \tag{5.168}$$

Following [58] we may now evaluate the lifetime ratio between charged and neutral B mesons in terms of B_i and ε_i,

$$\frac{\tau(B^-)}{\tau(B_d)} = 1 + k_1 B_1 + k_2 B_2 + k_3 \varepsilon_1 + k_4 \varepsilon_2 \,, \tag{5.169}$$

where the coefficients k_i are determined by the mass ratio z and the Wilson coefficients c_1 and c_2 (Table 5.8).

Table 5.8. Values of the coefficients k_i for different choices of the renormalization scale μ. The value of z is fixed at $z = 0.085$. The table is taken from [58]

μ	k_1	k_2	k_3	k_4
$m_b/2$	$+0.044$	0.003	-0.735	0.201
m_b	$+0.020$	0.004	-0.697	0.195
$2m_b$	-0.008	0.007	-0.665	0.189

In earlier work [59], the assumption of factorization was made and hence the matrix elements of the octet–octet operators ε_i vanish. Using this assumption, one obtains

$$\frac{\tau(B^-)}{\tau(B_d)} = 1 + 0.05 \times \frac{f_B^2}{(200 \text{ MeV})^2} \tag{5.170}$$

for the lifetime ratio, which is certainly compatible with the data. On the other hand, the coefficients k_3 and k_4 of the octett–octett operators are larger by a factor of five to ten, and if one expects these matrix elements to be suppressed by a factor $1/N_C$ it is a possible scenario that the non-factorizable contributions actually dominate the corrections to the lifetime ratio.

The lifetime of the Λ_b baryon has also been considered within the framework of the $1/m_Q$ expansion. Although naively the lifetime difference between the Λ_b baryon and the B meson is of order $1/m_b^2$, we may estimate the contributions of order $1/m_b^2$ to be small. For the Λ_b, we have $\lambda_2 = 0$ while the kinetic energy can be estimated by using [58]

$$[M(\Lambda_b) - M(\Lambda_c)] - [\bar{M}(B) - \bar{M}(D)] \\ = (\lambda_1(B) - \lambda_1(\Lambda_b)) \left(\frac{1}{m_c} - \frac{1}{m_b} \right) + \mathcal{O}(1/m_c^3) \tag{5.171}$$

where $\bar{M}(H) = (M(H_{0^-}) + 3M(H_{1^-}))/4$ is the spin-averaged mass of the mesons. From this estimate, we find that the $\mathcal{O}(1/m_b^2)$ contribution to the lifetime difference between Λ_b and B is as small as 2%. Thus the general conclusion of the discussion is that the lifetime ratio $\tau(\Lambda_b)/\tau(B_d)$ should be slightly smaller than unity, but cannot be as low as the measured value for "reasonable" values of the parameters. This fact constitutes a potential problem for the $1/m_Q$ expansion for inclusive non-leptonic decays, which is still a subject of research.

5.4.4 FCNC Decays of B Mesons

In this subsection we consider rare decays based on the quark transition $b \to s$, i.e. FCNC processes involving b quarks. This class of transitions has a rich phenomenology, which would fill a textbook on its own. However, the application of effective-field-theory methods can be demonstrated by focusing on a specific decay, which we chose to be the process $b \to s\gamma$. This process is currently the most interesting one, since on the one hand in a mature state as far as the corresponding theory is concerned, and on the other hand there are some data to compare the theory with.

Radiative rare B decays have attracted considerable attention in the last few years. After the first observation in 1994, by the CLEO collaboration [60], the data have become quite precise [61] so that even a measurement of the CP asymmetry in these decays [62] has become possible. As far as data are concerned, the situation clearly will improve further, after the excellent start of th two B factories at KEK and SLAC.

$B \to X_s\gamma$ tests the Standard Model in a particular way. Since there are no tree-level contributions to these processes in the Standard Model, these processes can occur only at the one-loop level. The GIM cancellation, which is present in all the FCNC processes, is lifted in this case by the large top-quark mass; if the top quark were as light as the b quark, these decays would be too rare to be observable.

Since the Standard Model contribution is small, these decays have a good sensitivity to "new physics", for example to new (heavy) particles contributing to the loop. In fact, already the first CLEO data could already constrain some models for "new physics" in a stringent way [60].

The effective Hamiltonian has already been discussed in Sect. 5.1. However, when calculating the amplitude, one has to calculate the matrix element of the effective Hamiltonian, leading to contributions from all operators with the correct quantum numbers; for example, there will be a matrix element of the operator O_2.

We shall first discuss the non-perturbative corrections to this decay. These arise from various sources, and we shall consider here only the subleading terms in the heavy-mass expansion, which are $1/m_b^2$ and $1/m_c^2$ corrections, and non-perturbative contributions to the photon energy spectrum, which are obtained from another application of the twist expansion and QCD factorization using SCET.

As far as the total rate is concerned, we have the subleading corrections of order $1/m_b^2$, which are parametrized in terms of the kinetic energy λ_1 and the chromomagnetic moment λ_2 defined in Sect. 4.3. In terms of these two parameters, the total rate reads, at tree level up to order $1/m_b^2$,

$$\Gamma = \frac{G_F^2 \alpha m_b^5}{32\pi^4} |V_{ts} V_{tb^*}|^2 |C_7|^2 \left(1 + \frac{\lambda_1 - 9\lambda_2}{2m_b^2} + \cdots \right) . \tag{5.172}$$

It is worth noting that if we assume that the charm quark is heavy too, we obtain non-perturbative contributions from the $1/m_c$ expansion also [63]. The relevant contribution originates from the four-fermion operators (e.g. the operator O_2) involving the charm quark; see Fig. 5.12a. Expanding the matrix element of O_2 in powers of $1/m_c$ we obtain a local operator of the form

$$O_{1/m_c^2} = \frac{1}{m_c^2} \bar{s} \gamma_\mu (1 - \gamma_5) T^a b \, G^a_{\nu\lambda} \epsilon^{\mu\nu\rho\sigma} \partial^\lambda F_{\rho\sigma} , \tag{5.173}$$

which can interfere with the leading term O_7 (see Fig. 5.12b).

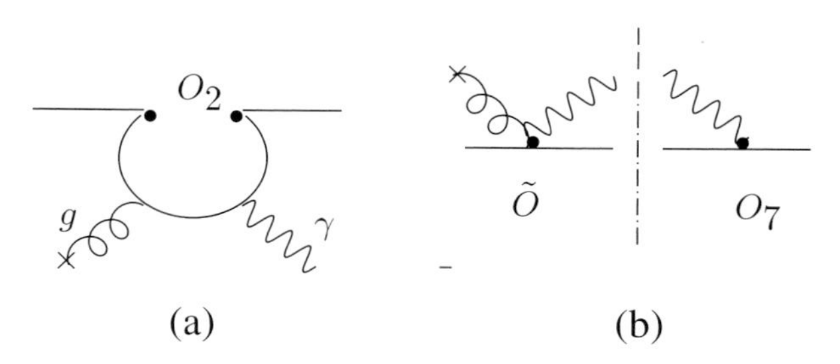

Fig. 5.12. Interference between O_7 and one of the four-fermion operators (O_2), leading to a contribution of order $1/m_c^2$

A detailed calculation [64, 65] reveals that this contribution is rather small,

$$\frac{\delta \Gamma_{1/m_c^2}}{\Gamma} = -\frac{C_2}{9C_7}\frac{\lambda_2}{m_c^2} \approx 0.03 \,, \tag{5.174}$$

and thus can safely be ignored at the current precision.

The main perturbative corrections are the QCD corrections, which are substantial. These corrections have been calculated using athe effective-field-theory framework described in Sect. 4.1. For the $b \to s\gamma$ transition, it turns out that the corrections at leading-logarithmic accuracy (which already involves a two-loop calculation) are substantial and, in addition, exhibit a sizeable dependence on the renormalization point. Thus the next-to-leading contributions are also known. The calculation involves the matching at subleading order

$$c_i(M_W) = c_i^{(0)}(M_W) + \frac{\alpha_s(M_W)}{\pi} c_i^{(1)}(M_W) + \cdots \tag{5.175}$$

and the calculation of the renormalization-group running using the anomalous-dimension matrix (5.71), including the next term in α_s, which we did not show in (5.71). Note that from the anomalous dimension matrix this involves a three-loop calculation.

The coefficient functions have the schematic form given in (4.19). Solving the renormalization group equations yields a resummation of the logarithms in the first and second columns of (4.19). Thus all terms of the form

$$\begin{aligned} c_i(\mu) = {} & c_i^{(0)}(M_W) \sum_{n=0} b_n^{(0)} \left(\frac{\alpha_s}{\pi} \ln \left(\frac{M_W^2}{\mu^2} \right) \right)^n \\ & + \frac{\alpha_s}{\pi} c_i^{(1)}(M_W) \sum_{n=0} b_n^{(1)} \left(\frac{\alpha_s}{\pi} \ln \left(\frac{M_W^2}{\mu^2} \right) \right)^n + \cdots \end{aligned} \tag{5.176}$$

are included.

The last step is to compute the matrix elements of the operators at a scale $\mu \approx m_b$. This can be done for the inclusive case using the $1/m_b$ expansion, while for the exclusive case the form factors are needed, which involves non-perturbative physics. Thus we shall focus on the inclusive case.

The leading and subleading terms of the coefficients have been calculated [66, 67, 68, 69], including electroweak contributions [70], the main part of which is due to the correct setting of the scale in α_{em}. A complete and up-to-date compilation can be found in [71]. Without going into any more detail, we shall quote the result from [71],

$$Br(B \to X_s\gamma) = (3.29 \pm 0.33) \times 10^{-4} \,, \tag{5.177}$$

where this result includes a cut on the photon energy at $E_{\gamma,min} = 0.05\, m_b$.

The QCD corrections are in fact dramatic; they increase the rate for $b \to s\gamma$ by about a factor of two. For example, even at the leading-log level we have $c_7(m_b)/c_7(M_W) = 1.63$. Another indication of this fact is a substantial dependence of the leading-order result on the choice of the renormalization scale μ. This is usually estimated by varying the scale μ between $m_b/2$ and $2m_b$. In this way we obtain a variation of $\delta_\mu = {}^{+27.4\%}_{-20.4\%}$ for the leading-order result. At subleading order this scale dependence is drastically reduced to [71] $\delta_\mu = {}^{+0.1\%}_{-3.2\%}$, which is actually smaller than one would expect.

However, it has recently been pointed out [72] that there is still a problem related to the definition of the charm quark mass appearing in the calculation of the matrix elements. In fact, at the two-loop level, one has contributions from diagrams such as the ones shown in Fig. 5.13.

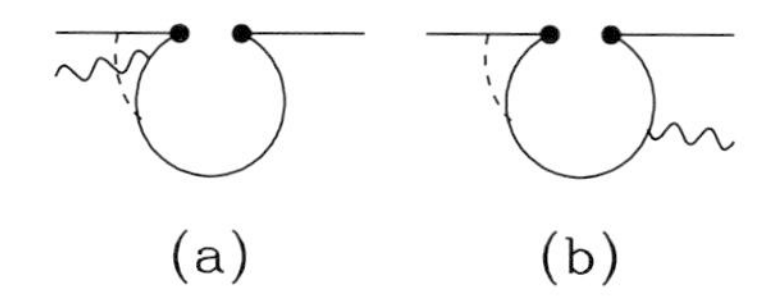

Fig. 5.13. Examples of diagrams leading to an m_c/m_b dependence at two loops

The main point is that the dependence on the parameter m_c/m_b is very steep, and small variations change the prediction for the branching fraction dramatically. It has been argued in [72] that one should use the pole mass for the b quark, since the b quark is an external line with the b quark in the B meson almost on-shell, while the c quark is inside a loop and hence a short-distance mass such as the $\overline{\text{MS}}$ mass is appropriate. Of course, this is only a guess for the higher-order corrections, but it indicates the range of the uncertainties.

Inserting $m_c^{\overline{\text{MS}}}(m_b)/m_b^{\text{pole}}$ shifts the prediction for the rate by one sigma upwards compared with the values obtained with $m_c^{\text{pole}}/m_b^{\text{pole}}$, and hence the uncertainties from this source are significant. Taking this as an estimate of the uncertainties induced by higher orders, we conclude that

$$\text{Br}(B \to X_s\gamma) = (3 - 4) \times 10^{-4}\,, \tag{5.178}$$

which indicates that at the current level of precision the usefulness of $B \to X_s\gamma$ is reduced. In order to settle this issue, a next-to-next-to-leading-order calculation will have to b performed, involving an anomalous dimension at the four-loop level and the calculation of the three-loop finite terms. Clearly, this is a technical challenge. The first steps have been performed, namely the n_f dependent terms have been investigated [73].

Finally we have to discuss the predictions for the photon energy spectrum. This spectrum is needed, because the extraction of the the process $B \to X_s\gamma$ requires one to introduce a lower cut on the photon energy to get rid of uninteresting processes such as ordinary bremsstrahlung. Clearly it is desirable to have this cut as high as possible, but this makes the process "less inclusive" and hence more sensitive to non-perturbative contributions to the photon energy spectrum.

Since we are dealing at tree level with a two-body decay, the naive calculation of the photon spectrum yields a δ-function at partonic level, and the $1/m_b^n$ corrections are again distributions located at the partonic energy $E_\gamma = m_b/2$, which are derivatives of δ-functions (see also Sect. 4.6). The result of such a calculation is at tree level yields a result, which has been given already in (4.96). Clearly, one cannot use such an expression to implement a cut on the photon energy spectrum, since this is not a smooth function.

The perturbative contributions have been calculated [67] and yield a spectrum that is determined mainly by the bremsstrahlung of a radiated gluon. This part of the calculation is fully perturbative and enters the next-to-leading-order analysis described above. In particular, the partonic δ function becomes smoother and turns into distributions of the form [74]

$$\frac{d\Gamma}{dx} = \cdots + \frac{\alpha_s}{\pi} \left[\left(\frac{\ln(1-x)}{1-x} \right)_+ , \left(\frac{1}{1-x} \right)_+ \right] , \tag{5.179}$$

where the ellipsis denotes terms that are regular as $x \to 1$ and contributions proportional to $\delta(1-x)$, which are determined by virtual gluons.

Here we shall focus on the nonperturbative contributions close to the endpoint. The general structure of the terms in the $1/m_b$ expansion is

$$\frac{d\Gamma}{dx} = \Gamma_0 \left[\sum_i a_i \left(\frac{1}{m_b} \right)^i \delta^{(i)}(1-x) + \mathcal{O}((1/m_b)^{i+1} \delta^{(i)}(1-x)) \right] , \tag{5.180}$$

showing that the leading terms have to be resummed, since they are all of the same size. This is in fact the master example of the twist expansion discussed in Sect. 4.6. The result is

$$\frac{d\Gamma}{dx} = \frac{G_F^2 \alpha m_b^5}{32\pi^4} |V_{ts} V_{tb^*}|^2 |C_7|^2 f(m_b[1-x]) , \tag{5.181}$$

where the light-cone distribution function f has been defined in (4.109).

Using the relations given in Sect. 4.6, we may also include subleading contributions. To this end, we have to compute the matching from full QCD to the twist expansion to obtain the Wilson-coefficient functions. We find for the subleading terms [75]

$$\begin{aligned} \frac{m_b}{\Gamma_0^s} \frac{d\Gamma}{dE_\gamma} &= (4E_\gamma - m_b) F(m_b - 2E_\gamma) \\ &+ \frac{1}{m_b} \left[h_1(m_b - 2E_\gamma) + H_2(m_b - 2E_\gamma) \right] , \end{aligned} \tag{5.182}$$

where

$$\Gamma_0^s = \frac{G_F^2 |V_{tb} V_{ts}^*|^2 \alpha |C_7^{\text{eff}}|^2 m_b^5}{32\pi^4} \tag{5.183}$$

and the functions F, H_2 and h_1 have been defined in (4.128), (4.127) and (4.122).

As has been pointed out earlier in this section, the relation between the photon spectrum for $B \to X_s\gamma$ and the lepton-energy spectrum in $B \to X_u \ell \bar{\nu}_\ell$ can be used to perform a model-independent extraction of V_{ub}, or, more precisely, V_{ub}/V_{ts}. However, the current data on the photon energy spectrum for $B \to X_s\gamma$ are not precise enough to be sensitive to the subleading shape functions.

5.4.5 Exclusive Non-Leptonic Decays

This class of decays is notoriously difficult to describe and effective field theory has been applied to these decays only recently [76, 77, 78, 79, 80, 81, 82, 83]. A model which has been frequently used in the past, mainly owing to the lack of any other method, is the so-called naive factorization. This approach estimates the matrix elements of four-quark operators by factorizing them into a product of two currents. As an example, consider the decays $B \to D\pi$, mediated at tree level by the operator

$$O_2 = \left(\bar{c}_{L,i}\gamma_\mu b_{L,i}\right)\left(\bar{d}_{L,j}\gamma_\mu u_{L,j}\right) \, . \tag{5.184}$$

In order to compute the rate, one has to calculate the matrix element of this operator between a B meson state and the $D\pi$ final state. Starting with the simplest case of a $\overline{B}^0 \to D^+\pi^-$ decay, the naive factorization in this case is

$$\langle D^+\pi^-|O_2|\overline{B}^0\rangle_{\text{fact}} = \langle D^+|\left(\bar{c}_{L,i}\gamma_\mu b_{L,i}\right)|\overline{B}^0\rangle\langle\pi^-|\left(\bar{d}_{L,j}\gamma_\mu u_{L,j}\right)|0\rangle \, . \tag{5.185}$$

Here we have introduced the subscript "fact" to identify the matrix elements which have been estimated by this approach.

Defining the form factors for the transition of a B meson into a pseudoscalar meson in the usual way,

$$\langle M(p')|\left(\bar{q}_{L,i}\gamma_\mu b_{L,i}\right)|B(p)\rangle = F_{B\to M}(q^2)(p_\mu + p'_\mu) + f_{B\to M}(q^2)q_\mu \, , \tag{5.186}$$

where $q = p - p'$ we obtain the following for the matrix element in naive factorization:

$$\langle D^+\pi^-|O_2|\overline{B}^0\rangle_{\text{fact}} = F_{B\to D}(M_\pi^2)(M_B^2 - M_D^2)f_\pi \, . \tag{5.187}$$

That is, the matrix element of the four quark operator is estimated to be the product of the $B \to D$ transition form factor at $q^2 = M_\pi^2$ and the pion decay constant. The relevant contribution is depicted in Fig. 5.14.

In the same way, this operator describes the decay $B^- \to D^0\pi^-$ as a product of a $B \to D$ form factor and the pion decay constant. $B^- \to D^0\pi^-$. However, in this case there is a second contribution, which is depicted in the second diagram of Fig. 5.15; here the roles of the two up quarks are interchanged.

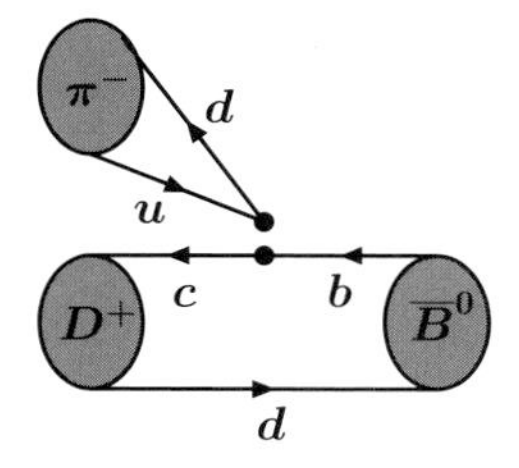

Fig. 5.14. Illustration of the factorized matrix element of O_2 for the transition $\overline{B}^0 \to D^+\pi^-$. The diagram has to be understood as a flavour flow diagram

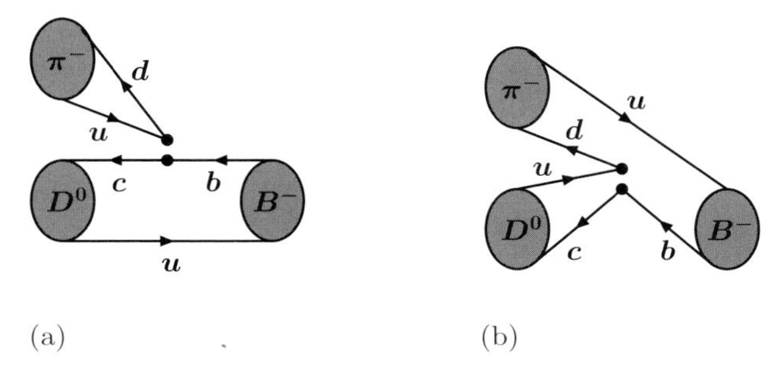

Fig. 5.15. Illustration of the two contributions to the factorized matrix element of O_2 in naive factorization for $B^- \to D^0\pi^-$

This second contribution can be obtained by first performing a Fierz transformation of the operator, which yields

$$O_2 = \left(\bar{d}_{L,j}\gamma_\mu b_{L,i}\right)\left(\bar{c}_{L,i}\gamma_\mu u_{L,j}\right) . \tag{5.188}$$

Factorizing this into two matrix elements yields

$$\begin{aligned}&\langle D^0\pi^-|O_2|B^-\rangle_{\text{fact}}\\ &= \text{Fig. 5.15a} + \langle\pi^-|\left(\bar{d}_{L,j}\gamma_\mu b_{L,i}\right)|B^-\rangle\langle D^0|\left(\bar{c}_{L,i}\gamma_\mu u_{L,j}\right)|0\rangle\end{aligned} \tag{5.189}$$

Note that the two currents can now be decomposed into a colour singlet piece and a colour octet piece, of which the colour octet cannot contribute, since the states are color-neutral. The appropriate colour factor is $1/N_c = 1/3$, and hence we obtain

$$\begin{aligned}&\langle\pi^-|\left(\bar{d}_{L,j}\gamma_\mu b_{L,i}\right)|B^-\rangle\langle D^0|\left(\bar{c}_{L,i}\gamma_\mu u_{L,j}\right)|0\rangle\\ &= \frac{1}{N_c}F_{B\to\pi}(M_D^2)(M_B^2 - M_\pi^2)f_D ,\end{aligned} \tag{5.190}$$

where f_D is the D meson decay constant defined analogously to the pion decay constant; see (4.157).

The prescription of naive factorization is to sum all the possible contributions. Thus, in total, we obtain

$$\begin{aligned} &\langle D^0\pi^-|O_2|B^-\rangle_{\rm fact} \\ &= F_{B\to D}(M_\pi^2)(M_B^2 - M_D^2)f_\pi - \frac{1}{N_{\rm c}}F_{B\to\pi}(M_D^2)(M_B^2 - M_\pi^2)f_D\,, \end{aligned} \tag{5.191}$$

where the minus sign between the two terms originates from the Pauli principle, since the amplitude has to be antisymmetric with respect to the interchange of the two $\bar{u}$ quarks in the final state. However, this so-called Pauli interference can in principle be constructive as well as destructive; see below.

Finally, the operator O_2 can also mediate the decay $\overline{B}^0 \to D^0\pi^0$ which is due to the diagram shown in Fig. 5.16; this diagram is the same as that of Fig. 5.15b with the spectator u quark replaced by a spectator d quark. The matrix element in naive factorization reads

$$\langle D^0\pi^0|O_2|\overline{B}^0\rangle_{\rm fact} = \frac{1}{N_{\rm c}}F_{B\to\pi}(M_D^2)(M_B^2 - M_\pi^2)f_D\,. \tag{5.192}$$

Note that for this decay, this is the only contribution which is suppressed by a colour factor $1/N_{\rm c}$.

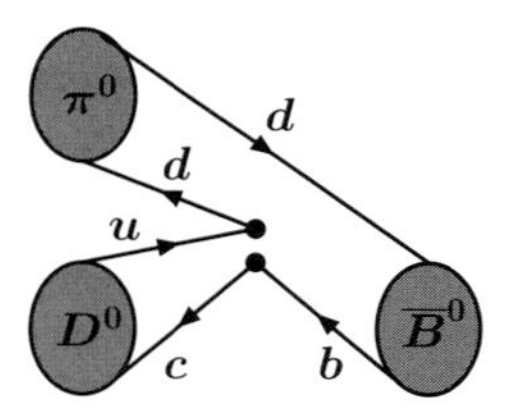

Fig. 5.16. Illustration of the factorized matrix element of O_2 for the transition $\overline{B}^0 \to D^0\pi^0$

The shortcomings of this ansatz are obvious. As was discussed in Sect. 5.1, the operator O_2 renormalizes; in particular, it mixes into the operator O_1 and, in more complicated cases, into a whole set of other operators. On the other hand, once naive factorization is performed, the matrix elements do not renormalize any more, since the left-handed currents do not have an anomalous dimension. This means that the two sides of (5.185) renormalize differently if the left-hand side is taken as the real QCD matrix element (i.e. without the subscript "fact"). In other words, the result of naive factorization depends on the scale at which it is performed.

Bauer, Stech and Wirbel (BSW) [84, 85] used this simple ansatz to define a model for non-leptonic two-body decays. They used the effective Hamiltonian as discussed in Sect. 5.1 and applied the above procedure to the matrix elements needed for a given transition. To circumvent the problem of the scale dependence, all the Wilson coefficients in the expressions for the effective Hamiltonian were replaced by phenomenological constants, which were assumed to be universal in the same way as the Wilson coefficients.

This model turns out to be surprisingly successful for non-leptonic two-body B decays. As an example, we again study the decay $\overline{B}^0 \to D^+\pi^-$, to which we have contributions from both O_1 and O_2 (see (5.22), (5.23) and (5.24)). Using naive factorization, we obtain

$$\langle D^+\pi^-|H_{\mathrm{eff}}|\overline{B}^0\rangle_{\mathrm{fact}} = i\frac{G_F}{\sqrt{2}}V_{cb}V_{ud}^*a_1F_{B\to D}(M_\pi^2)(M_B^2 - M_D^2)f_\pi \,, \tag{5.193}$$

where a_1 is a phenomenological constant. If we assume that naive factorization is actually valid at some scale μ_{f}, one obtains

$$a_1 = C_2(\mu_{\mathrm{f}}) + \frac{C_1(\mu_{\mathrm{f}})}{N_{\mathrm{c}}} \,, \tag{5.194}$$

where the second term comes from the Fierz-rearranged operator O_1. Note that the essence of the BSW model is that the parameter a_1 is treated as a universal quantity.

This type of transition is usually called a class I transition, since the current creating the meson from the vacuum is a charged current. All such decays are described by the universal parameter a_1 in the BSW model. Similarly, those transitions which have only a contribution from the Fierz-rearranged operators such that the current creating the meson from the vacuum is a neutral current are called class II transitions. One example has been considered above, which is the decay $\overline{B}^0 \to D^0\pi^0$, for which we obtain

$$\langle D^0\pi^0|H_{\mathrm{eff}}|\overline{B}^0\rangle_{\mathrm{fact}} = i\frac{G_F}{\sqrt{2}}V_{cb}V_{ud}^*a_2F_{B\to \pi}(M_D^2)(M_B^2 - M_\pi^2)f_D \,, \tag{5.195}$$

where a_2 is a universal parameter for these decays. If we again assume that naive factorization is actually valid at some scale μ_{f}, one obtains

$$a_2 = C_1(\mu_{\mathrm{f}}) + \frac{C_2(\mu_{\mathrm{f}})}{N_{\mathrm{c}}} \,. \tag{5.196}$$

Owing to the signs and the relative sizes of the Wilson coefficients a_2 is much more sensitive to the scale μ_{f} than is a_1.

Finally, there are class III decays which have contributions from both, a_1 and a_2. An example for such a decay is $B^- \to D^0\pi^-$, where the two contributions are shown in Fig. 5.15. In the BSW approach, one obtains

$$\begin{aligned}\langle D^0\pi^-|H_{\mathrm{eff}}|\overline{B}^-\rangle_{\mathrm{fact}} &= i\frac{G_F}{\sqrt{2}}V_{cb}V_{ud}^* \\ &\times \left[a_1F_{B\to D}(M_\pi^2)(M_B^2 - M_D^2)f_\pi - a_2F_{B\to\pi}(M_D^2)(M_B^2 - M_\pi^2)f_D\right] \,.\end{aligned} \tag{5.197}$$

Only this class of decays can be used to determine the relative phase of a_1 and a_2, i.e. one can determine from these decays whether the Pauli interference is constructive or destructive.

It is worthwhile to point out that naive factorization is actually valid in the formal limit $N_c \to \infty$. In particular, the off-diagonal entries in the anomalous-dimension matrix (5.25) behave like $1/N_c$ which solves the problem of scale dependence mentioned above.

The approach of naive factorization can be employed further to include the penguin operators, the matrix elements of which are then evaluated along the same lines. Furthermore, one may investigate decays into two-body final states with a vector and a pseudoscalar particle as well as decays into two vector particles , in a similar way [86, 87].

Recently this naive factorization has been put on a more sound theoretical basis by a technique called QCD factorization [81]. The main problem with naive factorization is that in full QCD, non-factorizable contributions such as that shown in Fig. 5.17b are present. In [81], all diagrams at order α_s (among which are the ones shown in Fig. 5.17) have been investigated, assuming that the kinematics of the outgoing quarks are such that those quarks can form the final-state mesons, i.e. the u and d quarks forming the pion are assumed to move collinearily. It turns out that in the infinite-mass limit for the B meson, all infrared contributions can be absorbed into the form factor and pion decay constant just as in naive factorization. In turn, all corrections violating factorization either are of order $\alpha_s(\mu_b)$ (where μ_b turns out to be given by $\mu_b^2 = m_b \Lambda_{\text{QCD}}$) or are suppressed by inverse powers of m_b.[2]

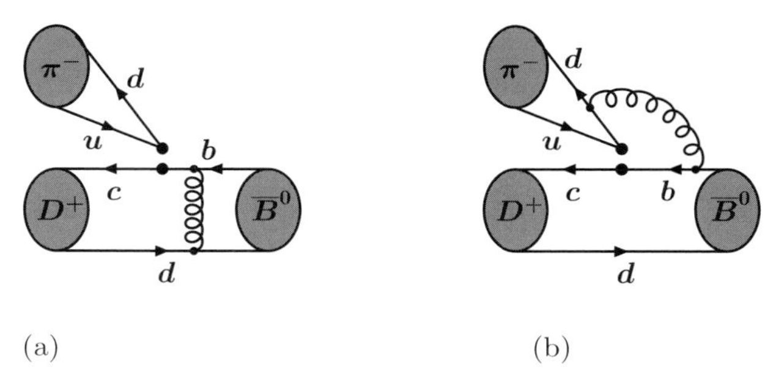

Fig. 5.17. QCD corrections to non-leptonic decays. (**a**) a factorizable contribution; (**b**) a non-factorizable contribution

Since QCD corrections can be computed systematically in QCD factorization, the problem of naive factorization concerning the scale dependence is not present. It has been shown in [81] that the scale dependence cancels properly order by order.

The systematics of this expansion and its relation to effective-field-theory approaches, in particular those approaches using a properly defined SCET,

[2]This statement is derived in [81] from perturbation theory. It could be that there are non-perturbative contributions which vanish more slowly than $1/m_b$.

are currently under investigation. In fact, SCET has been used to prove factorization to all orders in α_s for the case of the class I decay $\overline{B}^0 \to D^+\pi^-$. However, factorization for non-leptonic charmless decays has been proven up to now only on the basis of investigating the one-loop Feynman diagrams.

In parallel, the phenomenology of QCD factorization has been investigated. In particular, decays into two pseudoscalar particles [83] and two-body decays involving vector particles in the final state [88] have been investigated.

Exclusive non-leptonic decays not only are extremely relevant with respect to the branching ratio of each individual mode, they are also indispensable with respect to CP violation studies. Since both the form factors and the decay constants are real quantities in the usual convention, all the strong phases in the decays of B mesons will be either calculable perturbatively or suppressed by the large b quark mass, if QCD factorization holds. In this way, one can remove the most severe uncertainty in the evaluation of CP asymmetries of exclusive non-leptonic decays; we shall return to this point when considering CP violation in the next chapter.

References

1. F. J. Gilman and M. B. Wise, Phys. Rev. D **20**, 2392 (1979).
2. G. Buchalla, A. J. Buras and M. E. Lautenbacher, Rev. Mod. Phys. **68**, 1125 (1996) [arXiv:hep-ph/9512380].
3. M. A. Shifman, A. I. Vainshtein and V. I. Zakharov, Nucl. Phys. B **120**, 316 (1977).
4. M. A. Shifman, A. I. Vainshtein and V. I. Zakharov, Sov. Phys. JETP **45**, 670 (1977) [Zh. Eksp. Teor. Fiz. **72**, 1275 (1977)].
5. M. K. Gaillard and B. W. Lee, Phys. Rev. Lett. **33**, 108 (1974).
6. M. K. Gaillard and B. W. Lee, Phys. Rev. D **10**, 897 (1974).
7. G. Altarelli and L. Maiani, Phys. Lett. B **52**, 351 (1974).
8. B. Guberina and R. D. Peccei, Nucl. Phys. B **163**, 289 (1980).
9. R. Fleischer, Z. Phys. C **62**, 81 (1994).
10. B. Grinstein, R. P. Springer and M. B. Wise, Nucl. Phys. B **339**, 269 (1990).
11. B. Grinstein, M. J. Savage and M. B. Wise, Nucl. Phys. B **319**, 271 (1989).
12. T. Inami and C. S. Lim, Prog. Theor. Phys. **65**, 297 (1981); erratum, Prog. Theor. Phys. **65**, 1772 (1981).
13. M. Ciuchini, E. Franco, G. Martinelli, L. Reina and L. Silvestrini, Phys. Lett. B **316**, 127 (1993).
14. A. J. Buras, M. Misiak, M. Munz and S. Pokorski, Nucl. Phys. B **424**, 374 (1994) [arXiv:hep-ph/9311345].
15. W. S. Hou, R. S. Willey and A. Soni, Phys. Rev. Lett. **58**, 1608 (1987); erratum, Phys. Rev. Lett. **60**, 2337 (1988).
16. R. P. Feynman and M. Gell-Mann, Phys. Rev. **109**, 193 (1958).
17. K. Hagiwara et al. [Particle Data Group], Phys. Rev. D **66**, 010001 (2002).
18. M. L. Goldberger and S. B. Treiman, Phys. Rev. **111**, 354 (1958).
19. J. C. Hardy, I. S. Towner, V. T. Koslowsky, E. Hagberg and H. Schmeing, Nucl. Phys. A **509**, 429 (1990).

20. A. Sirlin, contribution to the International Workshop on Quark Mixing, CKM Unitarity, Heidelberg, 19–20 Sep. 2003.
21. H. Abele and D. Mund (eds.), Proceedings of the International Workshop on Quark Mixing, CKM Unitarity, Heidelberg, Germany, 19-20 Sep. 2002 arXiv:hep-ph/0312124.
22. H. Leutwyler and M. Roos, Z. Phys. C **25**, 91 (1984).
23. W. A. Bardeen, A. J. Buras and J. M. Gerard,
24. R. S. Chivukula, J. M. Flynn and H. Georgi,
25. J.E. Duboscq et al. (CLEO Collaboration), Phys. Rev. Lett. **76**, 3898 (1996).
26. M. E. Luke, Phys. Lett. B **252**, 447 (1990).
27. M. Battaglia et al. (eds.), Proceedings of the workshop on the CKM Matrix and the Unitarity Triangle, Geneva, 13–16 Feb. 2002 [arXiv:hep-ph/0304132].
28. I. Caprini and M. Neubert Phys. Lett. B **380**, 376 (1996).
29. C. G. Boyd, B. Grinstein and R. F. Lebed, Phys. Lett. B **353**, 306 (1995) [arXiv:hep-ph/9504235].
30. C. G. Boyd, B. Grinstein and R. F. Lebed, Phys. Rev. D **56**, 6895 (1997) [arXiv:hep-ph/9705252].
31. Heavy Flavor Averaging Group, http://www.slac.stanford.edu/xorg/hfag/
32. S. Hashimoto, A. S. Kronfeld, P. B. Mackenzie, S. M. Ryan and J. N. Simone, Phys. Rev. D **66**, 014503 (2002) [arXiv:hep-ph/0110253].
33. A. Ali and E. Pietarinen, Nucl. Phys. B **154**, 519 (1979).
34. N. Cabibbo, G. Corbo and L. Maiani, Nucl. Phys. B **155**, 83 (1979).
35. G. Altarelli et al., Nucl. Phys. B **208**, 365 (1982).
36. G. Corbo, Nucl. Phys. B **212**, 99 (1983).
37. M. Jezabek and J. H. Kühn, Nucl. Phys. B **320**, 20 (1989).
38. A. Falk et al., Phys. Rev. D **49**, 3367 (1994).
39. G. Altarelli et al., Nucl. Phys. B **208**, 365 (1982).
40. I. I. Y. Bigi, N. G. Uraltsev and A. I. Vainshtein, Phys. Lett. B **293**, 430 (1992); erratum, Phys. Lett. B **297**, 477 (1993) [arXiv:hep-ph/9207214].
41. I. I. Y. Bigi, M. A. Shifman, N. G. Uraltsev and A. I. Vainshtein, Phys. Rev. Lett. **71**, 496 (1993). [arXiv:hep-ph/9304225].
42. I. I. Y. Bigi, B. Blok, M. A. Shifman and A. I. Vainshtein, Phys. Lett. B **323** (1994) 408 [arXiv:hep-ph/9311339].
43. J. Chay, H. Georgi and B. Grinstein, Phys. Lett. B **247**, 399 (1990).
44. A. V. Manohar and M. B. Wise, Phys. Rev. D **49** 1310 (1994) [arXiv:hep-ph/9308246].
45. T. Mannel, Nucl. Phys. B **413**, 396 (1994) [arXiv:hep-ph/9308262].
46. T. Mannel and M. Neubert, Phys. Rev. D **50**, 2037 (1994) [arXiv:hep-ph/9402288].
47. M. Neubert, Phys. Lett. B **543**, 269 (2002) [arXiv:hep-ph/0207002].
48. C. W. Bauer, M. Luke and T. Mannel, Phys. Lett. B **543**, 261 (2002) [arXiv:hep-ph/0205150].
49. A. K. Leibovich, Z. Ligeti and M. B. Wise, Phys. Lett. B **539**, 242 (2002) [arXiv:hep-ph/0205148].
50. I. I. Y. Bigi, R. D. Dikeman and N. Uraltsev, Eur. Phys. J. C **4**, 453 (1998) [arXiv:hep-ph/9706520].
51. A. F. Falk, Z. Ligeti and M. B. Wise, Phys. Lett. B **406**, 225 (1997) [arXiv:hep-ph/9705235].

52. R. D. Dikeman and N. G. Uraltsev, Nucl. Phys. B **509**, 378 (1998) [arXiv:hep-ph/9703437].
53. G. Buchalla and G. Isidori, Nucl. Phys. B **525**, 333 (1998) [arXiv:hep-ph/9801456].
54. C. W. Bauer, Z. Ligeti and M. E. Luke, Phys. Rev. D **64**, 113004 (2001) [arXiv:hep-ph/0107074].
55. G. Korchemsky and G. Sterman Phys. Lett. B **340**, 96 (1994).
56. B. Guberina, S. Nussinov, R. D. Peccei and R. Ruckl, Phys. Lett. B **89**, 111 (1979).
57. N. Cabibbo and L. Maiani, Phys. Lett. **89B**, 111 (1979).
58. M. Neubert and C.T. Sachrajda, Nucl. Phys. **B 483**, 339 (1997).
59. I. I. Y. Bigi, B. Blok, M. A. Shifman, N. Uraltsev and A. I. Vainshtein, in *B Decays*, 2nd edn., p. 132, ed. S. Stone, (World Scientific 1994)
60. M. S. Alam *et al.* [CLEO Collaboration], Phys. Rev. Lett. **74**, 2885 (1995).
61. M. Nakao, contribution to the 21st International Symposium on Lepton and Photon Interactions at High Energies (LP 03), Batavia, Illinois, 11–16 Aug. 2003, arXiv:hep-ex/0312041.
62. T. E. Coan et al. [CLEO Collaboration], hep-ex/0010075.
63. M. B. Voloshin, Phys. Lett. B **397**, 275 (1997) [hep-ph/9612483].
64. G. Buchalla, G. Isidori and S. J. Rey, Nucl. Phys. B **511**, 594 (1998) [hep-ph/9705253].
65. Z. Ligeti, L. Randall and M. B. Wise, Phys. Lett. B **402**, 178 (1997) [hep-ph/9702322].
66. K. Chetyrkin, M. Misiak and M. Munz, Phys. Lett. B **400**, 206 (1997) [hep-ph/9612313].
67. A. Ali and C. Greub, Z. Phys. C **49**, 431 (1991).
68. K. Adel and Y. Yao, Phys. Rev. D **49**, 4945 (1994) [hep-ph/9308349].
69. C. Greub, T. Hurth and D. Wyler, Phys. Lett. B **380**, 385 (1996) [hep-ph/9602281], Phys. Rev. D **54**, 3350 (1996) [hep-ph/9603404].
70. A. Czarnecki and W. J. Marciano, Phys. Rev. Lett. **81**, 277 (1998) [arXiv:hep-ph/9804252].
71. A. L. Kagan and M. Neubert, Eur. Phys. J. C **7**, 5 (1999) [hep-ph/9805303].
72. P. Gambino and M. Misiak, Nucl. Phys. B **611**, 338 (2001) [arXiv:hep-ph/0104034].
73. K. Bieri, C. Greub and M. Steinhauser, Phys. Rev. D **67**, 114019 (2003) [arXiv:hep-ph/0302051].
74. T. Mannel and S. Recksiegel, Phys. Rev. D **60**, 114040 (1999) [arXiv:hep-ph/9904475].
75. C. W. Bauer, M. E. Luke and T. Mannel, Phys. Rev. D **68**, 094001 (2003) [arXiv:hep-ph/0102089].
76. C. W. Bauer, S. Fleming, D. Pirjol and I. W. Stewart, Phys. Rev. D **63** (2001) 114020 [arXiv:hep-ph/0011336].
77. C. W. Bauer, D. Pirjol and I. W. Stewart, Phys. Rev. D **65**, 054022 (2002) [arXiv:hep-ph/0109045].
78. M. Beneke, A. P. Chapovsky, M. Diehl and T. Feldmann, Nucl. Phys. B **643**, 431 (2002) [arXiv:hep-ph/0206152].
79. M. Beneke and T. Feldmann, Phys. Lett. B **553**, 267 (2003) [arXiv:hep-ph/0211358].

80. C. W. Bauer, D. Pirjol and I. W. Stewart, Phys. Rev. Lett. **87**, 201806 (2001) [arXiv:hep-ph/0107002].
81. M. Beneke, G. Buchalla, M. Neubert and C. T. Sachrajda, Phys. Rev. Lett. **83**, 1914 (1999) [arXiv:hep-ph/9905312].
82. M. Beneke, G. Buchalla, M. Neubert and C. T. Sachrajda, Nucl. Phys. B **591**, 313 (2000) [arXiv:hep-ph/0006124].
83. M. Beneke, G. Buchalla, M. Neubert and C. T. Sachrajda, Nucl. Phys. B **606**, 245 (2001) [arXiv:hep-ph/0104110].
84. M. Wirbel, B. Stech and M. Bauer, Z. Phys. C **29**, 637 (1985).
85. M. Bauer, B. Stech and M. Wirbel, Z. Phys. C **34**, 103 (1987).
86. G. Kramer and W. F. Palmer, Phys. Rev. D **45**, 193 (1992).
87. G. Kramer, T. Mannel and W. F. Palmer, Z. Phys. C **55**, 497 (1992).
88. M. Beneke and M. Neubert, Nucl. Phys. B **675**, 333 (2003) [arXiv:hep-ph/0308039].

6 Applications II: $\Delta F = 2$ Processes and CP Violation

In this chapter we discuss processes where flavour quantum numbers change by two units; these processes correspond mainly to the phenomenon of particle–antiparticle mixing. This phenomenon is related to CP violation owing to the fact that mixing can induce CP violation, and for that reason we discuss the application of effective field theories to these effects in a single chapter.

As far as the application of effective-field-theory methods is concerned, we have to deal mainly with the construction of the effective interaction mediating a $\Delta F = 2$ transition. However, to make quantitative predictions, we again have the problem of calculating hadronic matrix elements which is in general not possible. One exception may be the processes involving B mesons, for which one may use the heavy-mass expansion.

Concerning the application of effective field theory to CP violation, it is only very recently that some progress has been made, since significant CP asymmetries appear mainly in exclusive non-leptonic decays. As already discussed in previous chapters, this class of decays is still the most difficult one and is a subject of current research. In this book we shall concentrate on giving a few examples only. A very complete discussion of all aspects of CP violation can be found in a dedicated textbook [1].

After a discussion of how CP violation emerges in the Standard Model, we collect together some basic relations needed to describe the phenomena. Particle-antiparticle mixing is described in the Wigner–Weisskopf approximation described in Sect. 6.2. After considering the effects of mixing, we turn to CP violation, restricting ourselves to a few general remarks.

6.1 CP Symmetry in the Standard Model

In order to investigate CP violation, in general we have to first define the action of a CP transformation on the various fields. The charge conjugation transformation exchanges the roles of particles and antiparticles, while parity transformation is the inversion of the space coordinates.

Hermitian (or real) scalar fields describe particles which are their own antiparticles; furthermore, these fields are invariant under a parity transformation and thus are invariant under the combined transformations C and P.

Hermitian (or real) pseudoscalar fields are invariant under CP, but change their sign under P, so they will change their sign under the combined transformations C and P.

Complex scalar fields describe charged scalar particles, and thus there are particle–antiparticle pairs with equal but opposite charges. Under a CP transformation the particles turn into their antiparticles, while they remain invariant under P, since they are scalars, whereas a complex pseudoscalar field would change its sign.

The transformation properties of a spinor field can be defined by looking at currents. The electromagnetic current obviously changes its sign when one goes from the particle to the antiparticle:

$$J^{\mu}_{em} \xrightarrow{\mathrm{C}} -J^{\mu}_{em} \,. \tag{6.1}$$

If this current is due to an electron, we have to define the charge conjugation matrix C acting on the electron spinor ψ

$$\psi \xrightarrow{\mathrm{C}} \psi^{\mathrm{C}} = C\bar{\psi}^{T} \,, \tag{6.2}$$

where the the superscript T means the transposed spinor. The matrix C is defined in such a way that the Dirac matrices transform as

$$C^{-1}\gamma_{\mu}C = -\gamma_{\mu}^{T} \,, \tag{6.3}$$

from which (6.1) follows immediately. In order to make the electromagnetic interaction invariant under C, we assign C $= -1$ to the photon. Since the action

$$S_{em} = \int d^4x \, J^{\mu}_{em}(x)A^{\mu}(x) \tag{6.4}$$

is also invariant under P, the electromagnetic part of the Standard Model is CP-invariant.

The same argument applies for the weak neutral current, which also is invariant under a combined CP transformation. In order to show this, we have to take into account the fact that the CP transformation turns left-handed particles into right-handed antiparticles. Taking this together with the transformation properties of the scalar (i.e. the Higgs) sector of the Standard Model, we conclude that a possible CP violation can occur only in the charged currents.

In order to investigate the charged currents, we have to perform a charge conjugation transformation using (6.2) on a charged current. For the charged current of quarks (in the mass eigenbasis)

$$J^{\mu}_{\mathrm{cc}} = \bar{\mathcal{U}}_L \gamma_{\mu} V_{CKM} \mathcal{D}_L \,, \tag{6.5}$$

we obtain the following for the charge-conjugate current:

$$J^{\mu}_{\mathrm{cc}} \xrightarrow{\mathrm{C}} \bar{\mathcal{D}}_L V^{T}_{CKM} \gamma_{\mu} \mathcal{U}_L \,. \tag{6.6}$$

The charged-current contribution to the action is

$$S_{cc} = \int d^4x \left[\bar{\mathcal{U}}_L \gamma_\mu V_{CKM} \mathcal{D}_L W_\mu^+ + \bar{\mathcal{D}}_L V_{CKM}^\dagger \gamma_\mu \mathcal{U}_L W_\mu^- \right] \tag{6.7}$$

and hence we would have CP invariance if

$$V_{CKM}^\dagger = V_{CKM}^T \quad \text{or} \quad V_{CKM} = V_{CKM}^* \,, \tag{6.8}$$

i.e. for a real CKM matrix.

Obviously these arguments are somewhat simplified, since there is always the freedom to redefine the phases of the spinor fields. For a more detailed discussion we refer the reader to a textbook, such as [2]. In more general terms this means that CP violation can occur if there is no choice of the phases of the fields in which the CKM matrix can be made real. Note that the possibility of rephasing the quark fields has been already taken into account when we considered the CKM matrix in Sect. 3.2.

Thus in the Standard Model with three families and only a single Higgs doublet, the only source for CP violation is the single complex phase which remains even after using the freedom to rephase the fields. In other words, there is a coupling constant in the Lagrangian which has an "irreducible" phase, i.e. is complex. What remains to be discussed is how such a complex coupling leads to CP violation.

In practical terms CP violation means that there are observables which have different values if their CP images are considered. In the following we show that this always needs a CP-violating phase (e.g. a complex coupling) but also a non-trivial CP-conserving phase.

Any phase information can be obtained only by an interference experiment. Therefore we assume that an amplitude for some process consists of two contributions which can interfere; schematically, we obtaint

$$\mathcal{A} = \lambda_1 a_1 + \lambda_2 a_2 \,, \tag{6.9}$$

where we have extracted the complex couplings λ_1 and λ_2 explicitly, and a_1 and a_2 are matrix elements of the operators appearing in the (effective) Lagrangian. As an example, we may consider a weak decay of a particle: in this case λ_1 and λ_2 are combinations of CKM matrix elements, and a_1 and a_2 are usually hadronic matrix elements of quark currents.

From the amplitude, we compute the probability Γ (which, in the case of a particle decay, is the decay rate) and obtain

$$\Gamma = |\lambda_1|^2 |a_1|^2 + |\lambda_2|^2 |a_2|^2 + \mathrm{Re}(\lambda_1 \lambda_2^* a_1 a_2^*) \tag{6.10}$$

where the last term is the desired interference term. For the CP-conjugate process, we calculate the amplitude $\overline{\mathcal{A}}$. As we have seen above, the complex couplings turn into their complex conjugates, while a_1 and a_2 are CP-conserving matrix elements which do not change. Thus we obtain

$$\overline{\mathcal{A}} = \lambda_1^* a_1 + \lambda_2^* a_2 \ , \tag{6.11}$$

and the probability becomes

$$\overline{\Gamma} = |\lambda_1|^2|a_1|^2 + |\lambda_2|^2|a_2|^2 + \mathrm{Re}(\lambda_1^*\lambda_2 a_1 a_2^*) \ . \tag{6.12}$$

Clearly the interference term is different for the CP image; we define the CP asymmetry as

$$A_{CP} = \Gamma - \overline{\Gamma} = 2\,\mathrm{Im}[\lambda_1\lambda_2^*]\,\mathrm{Im}[a_1 a_2^*] \ , \tag{6.13}$$

which is non-vanishing only if we have an imaginary part of the couplings *and* a non-vanishing phase difference between the two contributions a_1 and a_2.

There are various origins of the CP-conserving phase difference. In the case of particle decays, it is in general a phase difference originating from strong interactions, but it can also be a phase difference originating from the time evolution, as it is the case for the time-dependent CP asymmetries to be discussed below.

6.2 $\Delta F = 2$ Processes: Particle–Antiparticle Mixing

Although the effect of particle–antiparticle mixing is not related to the phenomenon of CP violation, it is often mentioned in this context, and we shall give an outline of the effect here. The systems in which particle–antiparticle mixing has been observed are the K^0–$\overline{K}^0$ and the B^0–$\overline{B}^0$ systems; corresponding effects are expected in the B_s–$\overline{B}_s$ system.

We shall denote a (neutral) meson generically by H, i.e.

$$\begin{pmatrix} H \\ \overline{H} \end{pmatrix} = \begin{pmatrix} K^0 \\ \overline{K}^0 \end{pmatrix} \quad \text{or} \quad \begin{pmatrix} B^0 \\ \overline{B}^0 \end{pmatrix} \quad \text{or} \quad \begin{pmatrix} B_s \\ \overline{B}_s \end{pmatrix} \ , \tag{6.14}$$

and discuss the common features first.

Seond-order weak interactions can mediate transitions with $\Delta F = 2$, in which case we can have transitions between H and $\overline{H}$. A state prepared as a pure H state will develop a component of the state $\overline{H}$ after some time t, and thus we have to consider the time evolution of the system consisting of H and $\overline{H}$. Thus we have to consider the state

$$|\psi(t)\rangle = a(t)|H\rangle + \overline{a}(t)|\overline{H}\rangle \ , \tag{6.15}$$

with time dependent coefficients $a(t)$ and $\overline{a}(t)$. The time dependence of the coefficients is given by a Schrödinger equation

$$i\frac{d}{dt}\begin{pmatrix} a(t) \\ \overline{a}(t) \end{pmatrix} = \left[\mathcal{M} - \frac{i}{2}\Gamma\right]\begin{pmatrix} a(t) \\ \overline{a}(t) \end{pmatrix} \ , \tag{6.16}$$

where $\mathcal{M}$ and Γ are Hermitian 2×2 matrices. Note that the "Hamiltonian" appearing in (6.16) is non-Hermitian owing to the fact that H and $\overline{H}$ can decay and thus probability cannot be conserved in the simple 2×2 space spanned by H and $\overline{H}$.

In the rest frame of the meson H, the matrices $\mathcal{M}$ and Γ are given by

$$\left[\mathcal{M} - \frac{i}{2}\Gamma\right]_{ij} = m_H^{(0)}\delta_{ij} + \frac{1}{2m_H}\sum_n \frac{\langle H_i|H_{\text{weak}}|n\rangle\langle n|H_{\text{weak}}|H_j\rangle}{m_H^{(0)} - E_n + i\epsilon} + \cdots \,, \quad (6.17)$$

where the ellipsis denotes higher orders in H_{weak}, which we do not consider here. We use $H_1 \equiv H$ and $H_2 \equiv \overline{H}$ to simplify the notation.

The "absorptive piece" Γ is obtained by applying the identity

$$\frac{1}{\omega + i\epsilon} = \mathcal{P}\left(\frac{1}{\omega}\right) - i\pi\delta(\omega) \,, \quad (6.18)$$

where $\mathcal{P}()$ denotes the principal-value prescription yielding a real result. Thus we find

$$\Gamma_{ij} = \frac{1}{2m_H}\sum_n \langle H_i|H_{weak}|n\rangle\langle n|H_{weak}|H_j\rangle(2\pi)\delta(m_H^{(0)} - E_n) \quad (6.19)$$

for the absorptive piece. The diagonal entries are just the total width of H and $\overline{H}$.

Both $\mathcal{M}$ and Γ are Hermitian and so we have $\mathcal{M}_{12} = \mathcal{M}_{21}^*$ and $\mathcal{M}_{ii} = \mathcal{M}_{ii}^*$, as well as $\Gamma_{12} = \Gamma_{21}^*$ and $\Gamma_{ii} = \Gamma_{ii}^*$. Furthermore, CPT invariance requires the two diagonal elements to be equal, and thus the most general form is

$$\mathcal{M} - \frac{i}{2}\Gamma = \begin{pmatrix} \mathcal{M}_{11} - (i/2)\Gamma_{11} & \mathcal{M}_{12} - (i/2)\Gamma_{12} \\ \mathcal{M}_{12}^* - (i/2)\Gamma_{12}^* & \mathcal{M}_{11} - (i/2)\Gamma_{11} \end{pmatrix} = \begin{pmatrix} A & p^2 \\ q^2 & A \end{pmatrix} \,, \quad (6.20)$$

where A, p and q can be complex and

$$\frac{p}{q} = \sqrt{\frac{\mathcal{M}_{12} - (i/2)\Gamma_{12}}{\mathcal{M}_{12}^* - (i/2)\Gamma_{12}^*}} \,. \quad (6.21)$$

If CP were conserved, we would have $p = q$ and – using our freedom to chose the relative phase between H and $\overline{H}$ – we would end up with p and q being real. The eigenstates of the "Hamiltonian" $\mathcal{M} - (i/2)\Gamma$ would be the CP eigenstates K_1 and K_2 given in (5.107) and (5.108).

However, this is only true if the coupling constants appearing in the effective Hamiltonian are real in a suitable basis for the fields. If phases appear, for example through imaginary parts of the CKM factors, one cannot make these matrix elements real, and hence the eigenstates are not eigenstates of CP.

Even without CP invariance, we can still calculate the eigenstates of $\mathcal{M} - (i/2)\Gamma$, which become

$$|H_{\rm short}\rangle = \frac{1}{\sqrt{|p|^2+|q|^2}}\left(p|H\rangle - q|\overline{H}\rangle\right) \tag{6.22}$$

$$|H_{\rm long}\rangle = \frac{1}{\sqrt{|p|^2+|q|^2}}\left(p|H\rangle + q|\overline{H}\rangle\right) . \tag{6.23}$$

The difference of the two eigenvalues $m_{\rm short} - (i/2)\Gamma_{\rm short}$ and $m_{\rm long} - (i/2)\Gamma_{\rm long}$ is given by

$$\begin{aligned} 2pq &= (m_{\rm long} - m_{\rm short}) - \frac{i}{2}(\Gamma_{\rm long} - \Gamma_{\rm short}) \\ &= 2\sqrt{\left(\mathcal{M}_{12} - \frac{i}{2}\Gamma_{12}\right)\left(\mathcal{M}_{12}^* - \frac{i}{2}\Gamma_{12}^*\right)} . \end{aligned} \tag{6.24}$$

The solution of the time evolution (6.16) is given by

$$\begin{pmatrix} a(t) \\ \overline{a}(t) \end{pmatrix} = R(t) \cdot \begin{pmatrix} a(0) \\ \overline{a}(0) \end{pmatrix} , \quad \text{with} \quad R(t) = \begin{pmatrix} g_+(t) & (q/p)g_-(t) \\ (p/q)g_-(t) & g_+(t) \end{pmatrix} , \tag{6.25}$$

where

$$g_\pm = \frac{1}{2}\left(\exp\left[im_{\rm long}t - \frac{1}{2}\Gamma_{\rm long}t\right] \pm \exp\left[im_{\rm short}t - \frac{1}{2}\Gamma_{\rm short}t\right]\right) , \tag{6.26}$$

such that a state that is a pure H state at $t = 0$ evolves as

$$|\psi(t)\rangle = g_+(t)|H\rangle + \frac{p}{q}g_-(t)|\overline{H}\rangle . \tag{6.27}$$

The calculation of $\mathcal{M} - (i/2)\Gamma$ is generally not an easy task. Studying the relevant quark diagrams, we find that the $\Delta F = 2$ effects are induced by box diagrams of the type depicted in Fig. 6.1. In particular, the imaginary parts relevant to CP violation appear as a result of the irreducible phase in the CKM matrix. The question of which effects are phenomenologically relevant and of how to calculate or estimate them, depends on the particular system under consideration.

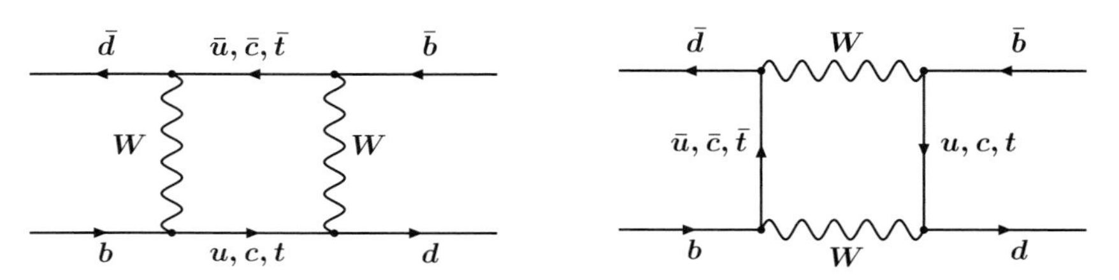

Fig. 6.1. Box diagrams mediating $\Delta F = \pm 2$ transitions. Only the case of a $B_0 \to \overline{B}_0$ transition is shown. For a $K_0 \to \overline{K}_0$ transitions the b quark has to be replaced by an s quark. Similarly, for a $B_s \to \overline{B}_s$ transition the d quark has to be replaced by an s quark

6.2.1 Mixing in the Kaon System

The kaon system is characterized by a large lifetime difference between the two neutral states. This has been discussed in Sect. 5.3: if CP were conserved, the K_{long} state would be a CP-odd state, which could not decay into two pions. This argument remains almost true, since CP violation is a small effect, i.e. the branching ratios for the CP-violating decays are small (see Sect. 6.3). Owing to the smaller phase space, the lifetimes differ in fact by a factor of about 580.

Since CP violation is small, we can write the states K_{long} and K_{short} in terms of CP eigenstates, where the small admixture of the "wrong" CP state is parametrized by a small (complex) quantity $\bar{\epsilon}$

$$|K_{\text{short}}\rangle = \frac{1}{\sqrt{1+|\bar{\epsilon}|^2}} \left(|K_1\rangle + \bar{\epsilon}|K_2\rangle \right) , \tag{6.28}$$

$$|K_{\text{long}}\rangle = \frac{1}{\sqrt{1+|\bar{\epsilon}|^2}} \left(|K_2\rangle + \bar{\epsilon}|K_1\rangle \right) , \tag{6.29}$$

where we have

$$\bar{\epsilon} = \frac{p-q}{p+q} = \frac{i}{2} \frac{\text{Im}\,\mathcal{M}_{12} - (i/2)\text{Im}\,\Gamma_{12}}{\text{Re}\,\mathcal{M}_{12} - (i/2)\text{Re}\,\Gamma_{12}} . \tag{6.30}$$

The mixing of the kaons can be discussed in the language of effective field theory. Constructing the effective interaction as described in Sect. 5.1 yields a $\Delta S = \pm 1$ effective Hamiltonian as given in (5.44). This effective Hamiltonian mediates transitions into final states which are common to both K_0 and $\overline{K}_0$. To second order in this effective Hamiltonian, we can have the process

$$K_0 \xrightarrow{H_{\text{eff}}} \left\{ \begin{array}{c} \pi^+\pi^- \\ \pi^0\pi^0 \\ \cdots \end{array} \right\} \xrightarrow{H_{\text{eff}}} \overline{K}_0 .$$

These transitions contribute to both $\mathcal{M}_{12}$ and Γ_{12} and are called long-distance contributions. They are very difficult to calculate for the case of kaons, which is one of the reasons why the parameters of Kaon CP violation are difficult to compute.

However, in addition to these long-distance contributions, we have (from the point of view of the scales involved in kaon decay) short-distance contributions which enter as effective interactions with $\Delta S = \pm 2$. These originate from the top- and charm-quark contributions appearing in the box diagrams depicted in Fig. 6.1. These contributions can be calculated, since the intermediate states can be treated as free quarks owing to their short-distance nature. Computing the box diagrams yields a $\Delta S = \pm 2$ contribution of the form

$$H_{eff}(\Delta S = \pm 2) = \frac{G_F^2}{16\pi^2}\left[(V_{td}^* V_{ts})^2 F(x_t) m_t^2 \eta_1 + (V_{cd}^* V_{cs})^2 F(x_c) m_c^2 \eta_2 \right.$$
$$\left. + 2(V_{td}^* V_{ts} V_{cd}^* V_{cs}) G(x_c, x_t) m_c^2 \eta_3 \right] \mathcal{O}_{\Delta S=2} + \text{h.c.} \,, \tag{6.31}$$

with a single operator mediating $\Delta S = 2$ transitions,

$$\mathcal{O}_{\Delta S=2} = (\bar{d}\gamma_\mu(1-\gamma_5)s)(\bar{d}\gamma^\mu(1-\gamma_5)s) \,. \tag{6.32}$$

In (6.31) $x_i = m_i^2/M_W^2$, and

$$F(x) = \frac{1}{4}\left[1 + \frac{9}{(1-x)} - \frac{6}{(1-x)^2} - \frac{6x^2}{(1-x)^3}\ln x\right] \,,$$
$$G(x,y) = \frac{y}{4}\left[\frac{\ln x}{x-y}\left(1 + \frac{6}{(1-x)} - \frac{3}{(1-x)^2}\right) + (x \leftrightarrow y) - \frac{6}{(1-x)(1-y)}\right] \tag{6.33}$$

are the relevant Inami–Lim functions [3]. The factors η_i are the short-distance QCD corrections calculated along the lines discussed in Sect. 5.1. These factors have been calculated in [4]; their values at leading logarithmic accuracy are

$$\eta_1 = 0.61 \qquad \eta_2 = 0.85 \qquad \eta_3 = 0.36 \,. \tag{6.34}$$

Including the known next-to-leading-order corrections, the values are [5, 6]

$$\eta_1 = 0.57 \qquad \eta_2 = 1.38 \qquad \eta_3 = 0.47 \,, \tag{6.35}$$

where we have quoted the so called renormalization scale and scheme independent coefficients. Note that the matrix elements have to be calculated using the same procedure; see below.

Note that the mass factors m_t^2 and m_c^2 indicate that the GIM mechanism is at work for these contributions. If all masses of the up-type quarks were equal, all the $\Delta S = \pm 2$ effects would vanish owing to an exact cancellation between the up, charm and top contributions. However, the masses of the up-type quarks are very different and thus a net effect remains. At first sight the top contribution seems enhanced by the large top mass; however, this contribution suffers from a substantial CKM suppression through the factor $(V_{td}^* V_{ts})^2 \sim \lambda^{10} \sim 2\times 10^{-7}$, which is sufficient to make the charm contribution the dominant one, since it overcompensates the relative factor $m_t^2/m_c^2 \sim 10^4$. The charm contribution is much less CKM suppressed, by only a factor $V_{cd}^* V_{cs} \sim \lambda \sim 0.2$. Taking into account the fact that $x_c \ll 1$ and $F(0) = 1$, we obtain approximately

$$H_{eff}(\Delta S = \pm 2) = \frac{G_F^2}{16\pi^2}(V_{cd}^* V_{cs})^2 m_c^2 \eta_2 \mathcal{O}_{\Delta S=2} + \text{h.c.} \tag{6.36}$$

Aside from the problem of the long-distance contributions, we still have to evaluate the matrix element of $\mathcal{O}_{\Delta S=2}$ between a K_0 and a $\overline{K}_0$ state to obtain the short-distance contribution to $\mathcal{M}_{12}$. This matrix element is usually estimated byuse of the naive factorization discussed in Sect. 5.4. In this approach, we have

$$\langle K_0|\mathcal{O}_{\Delta S=2}|\overline{K}_0\rangle = \frac{16}{3} f_K^2 m_K^2 B_K \,, \tag{6.37}$$

where we have introduced a so-called bag factor B_K, which is unity in naive factorization.

In general, the bag factor B_K depends on the renormalization scale and scheme. This dependence is unphysical and is compensated by the corresponding dependence of the short-distance QCD coefficients η_i.

It has become customary to quote the so-called renormalization-scale- and scheme-independent bag factor B_K, which at leading-log level is given by

$$B_K = [\alpha_s(\mu)]^{-2/9} B_K(\mu) \,. \tag{6.38}$$

Including the next-to-leading-order terms, the following result is obtained from lattice calculations [7]:

$$B_K \sim 0.87(6)(13) \,. \tag{6.39}$$

6.2.2 Mixing in the B_0-Meson System

The situation in the B system allows more precise estimates of the mixing parameters and of the CP violation induced by this effect. The main advantage is that the $\Delta B = \pm 2$ contributions are dominated by the top quark, since its CKM suppression through the factor $(V_{td}V_{tb}^*)^2 \sim \lambda^6$ is much weaker. Furthermore, the CKM factors of the other contributions are comparable, and consequently the large top-quark mass wins in the case of the B mesons. Thus the main conclusion is that the mixing of the B mesons is dominated by short-distance contributions.

Furthermore, the lifetime differences in the B_d meson system are small, since the absorptive part Γ_{12} is strongly CKM suppressed, such that one may equate Γ_{long} to Γ_{short} in the formula (6.26) for the time evolution of the neutral B_d states. The lifetime differences can be calculated using the heavy-mass expansion and are known even to next-to-leading-logarithmic accuracy [8].

For the B_s mesons, the lifetime differences have been estimated to be larger than in the B_d system [8], but we shall not discuss this matter here, since still the lifetime differences are still relatively small, roughly $\Delta\Gamma/\Gamma \sim$ 10%.

If we neglect the lifetime differences, the formulae for the time evolution simplify considerably. The 2×2 matrix describing the time evolution simplifies to

$$R(t) = e^{-iMt-(1/2)\Gamma t} \begin{pmatrix} \cos\left(\frac{1}{2}\Delta m\, t\right) & \frac{q}{p} i \sin\left(\frac{1}{2}\Delta m\, t\right) \\ -\frac{q}{p} i \sin\left(\frac{1}{2}\Delta m\, t\right) & \cos\left(\frac{1}{2}\Delta m\, t\right) \end{pmatrix} \tag{6.40}$$

where we have defined the average mass $M = (m_{\text{long}} + m_{\text{short}})/2$ and the mass difference $\Delta m = m_{\text{long}} - m_{\text{short}}$.

Furthermore, as $M_{12} \gg \Gamma_{12}$, the ratio p/q becomes just a phase,

$$\frac{p}{q} = \sqrt{\frac{\mathcal{M}_{12}}{\mathcal{M}_{12}^*}} = \exp(i\Phi_M) \tag{6.41}$$

where Φ_M is the weak mixing phase, coming from the CKM factors of the $\Delta B = 2$ effective Hamiltonian. Thus the final result for the time evolution is

$$R(t) = e^{-iMt-(1/2)\Gamma t} \begin{pmatrix} \cos\left(\frac{1}{2}\Delta m\, t\right) & ie^{-i\Phi_M} \sin\left(\frac{1}{2}\Delta m\, t\right) \\ -ie^{i\Phi_M} i \sin\left(\frac{1}{2}\Delta m\, t\right) & \cos\left(\frac{1}{2}\Delta m\, t\right) \end{pmatrix} . \tag{6.42}$$

The quantity Δm can be computed in the effective-field-theory approach, since the it is dominated by short-distance effects, namely those of the W boson and the top quark. Evaluating the box diagrams shown in Fig. 6.1 we obtains the following for the mixing in the B_d system:

$$H_{eff}(\Delta B = \pm 2) = \frac{G_F^2}{16\pi^2} \left[(V_{td}^* V_{tb})^2 F(x_t) m_t^2 \eta_1' \mathcal{O}_{\Delta B=2} + \text{h.c.} \right] , \tag{6.43}$$

with the operator

$$\mathcal{O}_{\Delta B=2} = (\bar{d}\gamma_\mu(1-\gamma_5)b)(\bar{d}\gamma^\mu(1-\gamma_5)b) ; \tag{6.44}$$

the function $F(x_t)$ is the same as for the $\Delta S = 2$ Hamiltonian. The QCD coefficient η_1' depends in general on the scheme and scale of renormalization. Using the so-called scale- and scheme-independent definition of this parameter, we obtain, at next-to-leading order accuracy [5],

$$\eta_1 = 0.551 . \tag{6.45}$$

In order to obtain the mass difference one has to calculate the matrix elements of the effective Hamiltonian between a B and a $\overline{B}$ state. One finds

$$\Delta m = \frac{G_F^2}{8\pi^2} (V_{td}^* V_{tb})^2 F(x_t) m_t^2 \eta_1' \langle B | \mathcal{O}_{\Delta B=2} | \overline{B} \rangle . \tag{6.46}$$

The remaining task is to evaluate the matrix element of the local operator $\mathcal{O}_{\Delta B=2}$ which contains the necessary non-perturbative information. It has become customary to parametrize this matrix element in the same way as in the kaon system by

$$\langle B_0|\mathcal{O}_{\Delta B=2}|\overline{B}_0\rangle = \frac{16}{3} f_B^2 m_B^2 B_B \;, \tag{6.47}$$

where the "bag factor" B_B is unity in naive factorization. In the scheme- and scale-independent definition (which has been used already for the QCD coefficient η_1'),

$$B_B = [\alpha_s(\mu)]^{-6/23} B_B(\mu) \;. \tag{6.48}$$

A numerical value can be obtained from lattice simulation, and the result obtained is [9]

$$B_B = \begin{cases} 230 \pm 30\,\mathrm{MeV} & \text{for the } B_s \\ 189 \pm 30\,\mathrm{MeV} & \text{for the } B_d \end{cases} . \tag{6.49}$$

In order to compute a number for Δm, we still need an input for the B meson decay constant, which has not been measured yet. Thus we have to rely on lattice simulations for this quantity also, and the result obtained is

$$f_B = \begin{cases} 1.34 \pm 0.10 & \text{for the } B_s \\ 1.30 \pm 0.12 & \text{for the } B_d \end{cases} . \tag{6.50}$$

This calculation has been used to obtain a value for the CKM matrix element V_{td}; the result is [10]

$$|V_{td}| = 0.0079 \pm 0.0015 \;. \tag{6.51}$$

We may consider B_s oscillations in the same way: we have the same relations except that, in the effective Hamiltonian, V_{td} is replaced by V_{ts} and the d quark is replaced by an s quark. Thus the oscillation frequency in the B_s system is much larger than in the B_d system, roughly by a factor [10]

$$\frac{\Delta m_s}{\Delta m_d} = \left|\frac{V_{ts}}{V_{td}}\right|^2 \sim 100 \;. \tag{6.52}$$

Futhermore, the mixing phase Φ_M introduced in (6.41) is, in the standard convention,

$$\Phi_M = \begin{cases} 2\beta & \text{for } B_d \\ 2\delta\gamma & \text{for } B_s \end{cases} \tag{6.53}$$

where $\delta\gamma$ is small, of the order of λ^6.

6.2.3 Mixing in the D_0-Meson System

The situation in the neutral D meson system is very different owing to the fact that the roles of up-type and down-type quarks in Fig. 6.1 are reversed: the external quarks are the valence quarks c and $\bar{u}$ of a neutral D meson, while the sum over the internal quarks runs over the down-type quarks d, s and b. Since the differences in masses for the down-type quarks are much smaller than the differences for the up-type quarks (which are mainly driven by the top quark), the GIM mechanism works much more efficiently than for the kaons and B mesons. Furthermore, owing to the small CKM angles the coupling to the third family is negligible, such that effectively only the first two generations play a role, making the GIM mechanism even more efficient.

In comparison with the cases of K–$\overline{K}$ and B–$\overline{B}$ mixing, the main contribution to D–$\overline{D}$ mixing is mainly from long-distance effects, involving processes such as

$$D_0 \xrightarrow{H_{eff}} \left\{ \begin{array}{c} K^+K^- \\ K^0\overline{K}^0 \\ \dots \end{array} \right\} \xrightarrow{H_{eff}} \overline{D}_0 \, , \tag{6.54}$$

which are again difficult to calculate. However, assuming that the charm mass m_c is still a perturbative scale, one may apply heavy-quark effective theory to calculate the corresponding matrix elements [11, 12, 13].

The operators of leading dimensionality are operators of dimension six. Calculating the matching obtained from the diagrams shown in Fig. 6.2, we find

$$\begin{aligned} H^{\Delta c=2}_{\mathrm{eff,a}} &= \frac{G_F^2}{2\pi^2} \sin^2\theta_C \cos^2\theta_C \frac{m_s^4}{m_c^2} \\ &\quad \times \left[2(\bar{u}_L\gamma_\mu c_v)(\bar{u}_L\gamma^\mu c_{-v}) + 4(\bar{u}_L c_v)(\bar{u}_L c_{-v}) \right] \end{aligned} \tag{6.55}$$

The GIM mechanism predicts that the result for D–$\overline{D}$ mixing has to be proportional to m_s^2 when the mass of the down quark is neglected. However, the loop diagram of Fig. 6.2 yields a suppression by another factor of m_s^2, so the total contribution is of the order of m_s^4.

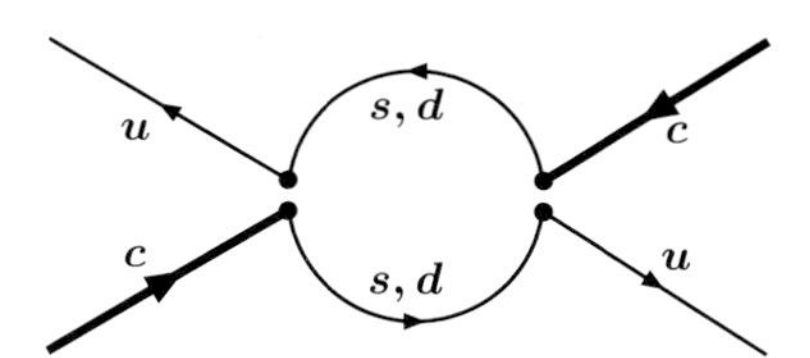

Fig. 6.2. One-loop contribution to the matching to the dimension-six operator for D–$\overline{D}$ mixing

This is in contradiction to calculations where the intermediate states shown in (6.54) are studied explicitly. The result is obtained, model independently, that each channel may be combined with its counterpart where $s \leftrightarrow d$, yielding a result proportional to m_s^2. From the effective-field-theory calculation, one would conclude that the amplitudes of the individual channels have to conspire in such a way that the leading term proportional to m_s^2 cancels, leaving the result (6.55).

The next order in the operator product expansion is the six-quark operators, which are obtained from Fig. 6.3. The contribution to the effective Hamiltonian reads schematically

$$H_{\mathrm{eff,b}}^{\Delta c=2} = G_F^2 \sin\theta_C \cos\theta_C \frac{m_s^2}{m_c^3} (\bar{u}_L \Gamma_1 c_v)(\bar{u}_L \Gamma_2 c_{-v})(\bar{d}_L \Gamma_3 d) \;, \qquad (6.56)$$

where the Γ_i are sums of combinations of Dirac matrices. Note that (6.56) is of the expected order in m_s.

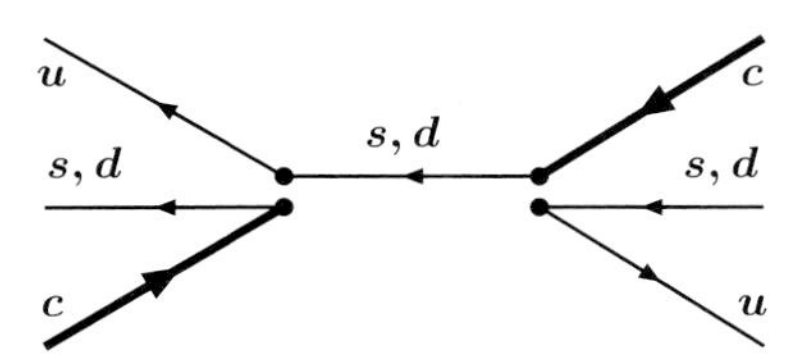

Fig. 6.3. Contribution to the matching to the dimension-nine (six-quark) operators for D–$\overline{D}$ mixing

Including also the eight-quark operators, the QCD renormalization properties of these operators have been studied at one loop in [13]. The results in that paper exclude the possibility that QCD effects are large and overcome the suppression by the factors m_s/m_c of these subleading terms, so as to make them the leading contribution. The result quoted for Δm_c in [13] is

$$\Delta m_c = (0.9 - 3.7) \times 10^{-17}\mathrm{GeV} \;, \qquad (6.57)$$

where the uncertainty is due to the estimates of the hadronic matrix elements.

The contributions of the third family to D–$\overline{D}$ mixing is extremely small and can be neglected. Thus D–$\overline{D}$ mixing (and in fact all charm physics) happens in the first two families. This has the consequence that CP-violating effects are practically absent; in particular, the mixing phase in D–$\overline{D}$ mixing is extremely small, leaving only a ridiculously small Standard Model contribution to mixing-induced CP violation in the D system.

6.3 Phenomenology of CP Violation: Kaons

The first, and very unexpected, manifestation of a violation of CP symmetry was found through non-leptonic decays in the system of neutral kaons. The phenomenology of these decays has been discussed already in Sect. 5.3. In that section we introduced the CP eigenstates of neutral kaons K_1 and K_2 in (5.107) and (5.108). If CP were a good symmetry, the long-lived kaon state would be identical to K_2 and could not decay into two pions. However, it was the surprising observation of Christensen, Cronin, Fitch and Turlay [14] that this decay indeed happens; the current values for the branching ratios are [10]

$$\mathrm{Br}(K_{long} \to \pi^+\pi^-) = (2.056 \pm 0.033) \times 10^{-3} \,, \tag{6.58}$$

$$\mathrm{Br}(K_{long} \to \pi^0\pi^0) = (9.27 \pm 0.19) \times 10^{-4} \,, \tag{6.59}$$

and clearly this effect is well established.

As already mentioned in Sect. 5.3, the two iso-doublet kaon states are connected through a CP transformation (see (5.105)). The decay amplitudes for the $K \to \pi\pi$ decays were discussed for the K^+ and K^0, where we introduced the isospin amplitudes A_0 and A_2. Analogously, we introduce the CP images of these amplitudes as

$$\langle \pi^-\pi^0 | H_{eff} | K^- \rangle = +\sqrt{\frac{3}{2}}\,\overline{A}_2 \,, \tag{6.60}$$

$$\langle \pi^0\pi^0 | H_{eff} | \overline{K}^0 \rangle = -\sqrt{\frac{1}{3}}\,\overline{A}_0 + \sqrt{\frac{2}{3}}\,\overline{A}_2 \,, \tag{6.61}$$

$$\langle \pi^+\pi^- | H_{eff} | \overline{K}^0 \rangle = +\sqrt{\frac{2}{3}}\,\overline{A}_0 + \sqrt{\frac{1}{3}}\,\overline{A}_2 \,. \tag{6.62}$$

If CP were conserved, we would have $\overline{A}_0 = A_0$ and $\overline{A}_2 = A_2$. Using the definitions of the eigenvectors K_{short} and K_{long}, we obtain the following for the CP-violating decays $K_{\mathrm{long}} \to \pi\pi$:

$$A(K_{\mathrm{long}} \to \pi^0\pi^0) = -\sqrt{\frac{1}{3}}\,\frac{pA_0 - q\overline{A}_0}{\sqrt{|p|^2+|q|^2}} + \sqrt{\frac{2}{3}}\,\frac{pA_2 - q\overline{A}_2}{\sqrt{|p|^2+|q|^2}} \,, \tag{6.63}$$

$$A(K_{\mathrm{long}} \to \pi^+\pi^-) = \sqrt{\frac{2}{3}}\,\frac{pA_0 - q\overline{A}_0}{\sqrt{|p|^2+|q|^2}} + \sqrt{\frac{1}{3}}\,\frac{pA_2 - q\overline{A}_2}{\sqrt{|p|^2+|q|^2}} \,. \tag{6.64}$$

It has become customary to consider the amplitude ratios

$$\eta_{+-} = \frac{\langle \pi^+\pi^- | H_{eff} | K_{long} \rangle}{\langle \pi^+\pi^- | H_{eff} | K_{short} \rangle} = \varepsilon + \varepsilon' \,, \tag{6.65}$$

$$\eta_{00} = \frac{\langle \pi^0\pi^0 | H_{eff} | K_{long} \rangle}{\langle \pi^0\pi^0 | H_{eff} | K_{short} \rangle} = \varepsilon - 2\varepsilon' \,, \tag{6.66}$$

which vanish in the limit of CP conservation.

Let us first study the simplified case in which we neglect the amplitude A_2 in comparison with A_0, which is a reasonable approximation in view of the $\Delta I = 1/2$ rule. Approximately, we can set $A_2 \approx 0$ and $\overline{A}_2 \approx 0$, and we obtain the following for the amplitude ratios:

$$\eta_{+-} = \eta_{00} = \varepsilon = \frac{1}{\sqrt{2}}\left(1 - \frac{q}{p}\frac{\overline{A}_0}{A_0}\right) \quad \text{and} \quad \varepsilon' = 0 \,. \tag{6.67}$$

A non-vanishing ε' can only appear if we have a non-vanishing A_2. Thus, owing to the $\Delta I = 1/2$ rule, we expect ε' to be much smaller than ε. In fact, if we keep the A_2 amplitudes, we find

$$\varepsilon' = \frac{1}{2}\frac{q}{p}\frac{A_2}{A_0}\left(\frac{\overline{A}_0}{A_0} - \frac{\overline{A}_2}{A_2}\right) \tag{6.68}$$

while the result for ε remains the same even in the presence of a non-vanishing A_2.

These results have a simple interpretation. A non-vanishing value of ε can appear even if the decay amplitudes are CP-conserving, in which case we have $A_0 = \overline{A}_0$. Then we have to have $p \neq q$ and thus CP violation originates from the interference of the two amplitudes $A(K^0 \to \pi\pi)$ and $A(K^0 \to \overline{K}^0 \to \pi\pi)$. Clearly this is an effect of kaon mixing and hence is the same for both two-pion final states. In fact, a non-vanishing value of ε' has been established only recently (see below), and before this discovery it was a realistic scenario that the CP violation in the kaon sector (which at that time was the only CP violation that had been observed) originated from an effect beyond the standard model. In such a "superweak" model [15] the CKM matrix may be assumed to be real, while the observed CP violation is due to a complex coupling in the $\Delta S = 2$ effective Hamiltonian originating from an effect beyond the Standard Model.

Similarly, ε' needs the presence of the A_2 amplitude, and CP violation is induced by an interference between the two contributions with different isospin, i.e. an interference of A_0 and A_2.

Experimentally, a non-vanishing value is found for both parameters ε and ε', where the latter rules out the superweak scenario. The Particle Data Group quotes

$$|\varepsilon| = (2.282 \pm 0.017) \times 10^{-3}\,, \quad \mathrm{Re}\frac{\varepsilon'}{\varepsilon} \approx \frac{\varepsilon'}{\varepsilon} = (1.8 \pm 0.4) \times 10^{-3}\,. \tag{6.69}$$

A theoretical prediction of these parameters is very difficult; one may consider the various quark diagrams contributing to the amplitudes to relate the two parameters ϵ and ϵ' to the phases in the CKM matrix. However, this is quite complicated for the kaon system, owing to our inability to calculate hadronic matrix elements reliably. Various groups have given estimates of ε', based on the effective Hamiltonian discussed in Sect. 5.1. A review on this subject has been given recently in [16]. In Fig. 6.4, the theoretical predictions of the various groups are compared to the experimental values.

6.4 Phenomenology of CP Violation: B Mesons

In this section we shall discuss some general aspects of the phenomenology of CP violation in the B meson system. While kaons have very few decay

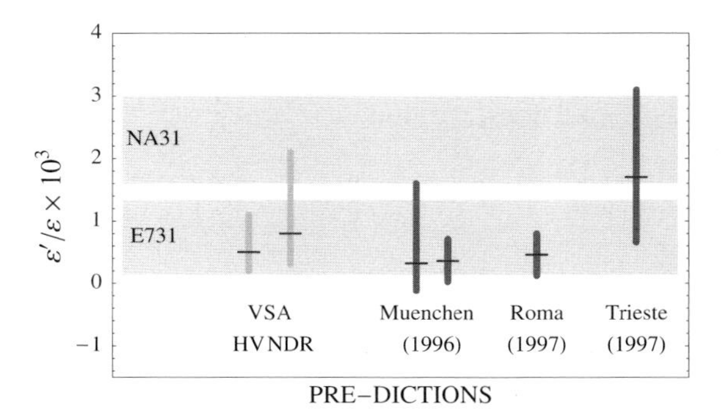

Fig. 6.4. Predictions of ε'/ε versus the data. "VSA" is the "vacuum saturation approximation", corresponding to naive factorization in two different schemes. "Muenchen" means [17], "Roma" means [18] and "Trieste" means [19]. The figure is taken from [20]

channels, B mesons can decay into many channels owing to their large mass. Unlike kaons, where we have only two relevant non-leptonic decay modes, B mesons have many non-leptonic decay modes, which yield a very rich CP phenomenology. Clearly, a complete description is beyond the scope of this book; furthermore, effective field theories are of only limited use here, since we are dealing mainly with non-leptonic decays. For a complete discussion of CP phenomenology, we refer the reader to a monography dealing exclusively with CP violation [1].

The situation in the system of B mesons is also different owing to the fact that the lifetime difference between the two eigenstates of B mesons is small. Clearly, this has an impact on how a CP asymmetry is measured in the system of B mesons.

We shall start our discussion with a collection of general relations. As already discussed in Sect. 6.1, a measurement of CP violation in general requires the interference of two amplitudes which have both a strong and a weak phase difference. In B decays this means that one can have a CP asymmetry

$$\mathcal{A}_{CP}(B^+ \to f) = \frac{\Gamma(B^+ \to f) - \Gamma(B^- \to \overline{f})}{\Gamma(B^+ \to f) + \Gamma(B^- \to \overline{f})} \tag{6.70}$$

in decays a charged B meson into a final state f. Such an effect is called *direct* CP violation and can also happen in neutral-B-meson decays. However, due to B–$\overline{B}$ oscillations, the situation in the system of neutral B mesons is more complicated.

The weak phases are, in the Standard Model, due to the CKM matrix elements; as an example we consider the decays $B \to K\pi$. In these decays, we have two possible contributions, shown in Fig. 6.5.

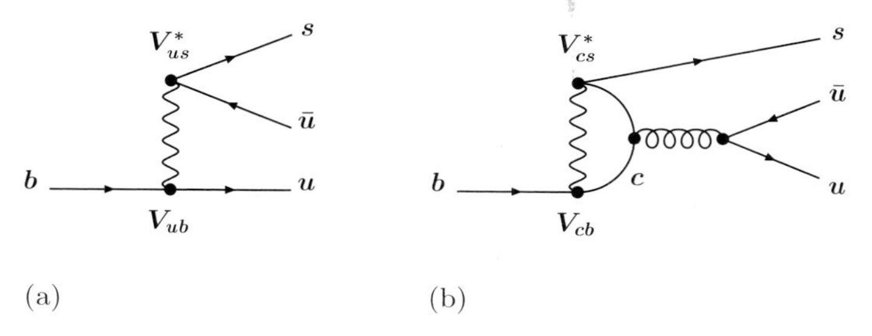

Fig. 6.5. Tree **a** and penguin **b** contributions to the decay $B \to K\pi$

Looking at the corresponding matrix elements, we find the following for the two decays $B_d \to K^+\pi^-$ and $B^+ \to K^0\pi^+$:

$$\begin{aligned} \langle \pi^- K^+ | H_{eff} | B_d \rangle &= -(P+T) \,, \\ \langle \pi^+ K^0 | H_{eff} | B_+ \rangle &= P \,, \end{aligned} \tag{6.71}$$

where P is the amplitude corresponding to "QCD penguin-like diagrams" (Fig. 6.5b) and T corresponds to the "tree amplitude" (Fig. 6.5a). The electroweak penguin contributions are colour suppressed, and we shall neglect them in the following.

When discussing CP asymmetries, we have to consider the CP image of the processes considered in (6.71). To this end, we have to identify the weak phases appearing in the amplitudes (6.71). Clearly the tree amplitude carries a weak phase factor $\exp(-i\gamma)$, while the CKM factors of the penguin amplitude are real in the usual convention.

From this we obtain

$$\frac{\mathrm{Br}(B_d \to \pi^- K^+)}{\mathrm{Br}(B^+ \to \pi^+ K^0)} = 1 + r^2 - 2r\cos(\delta + \gamma) \,, \tag{6.72}$$

$$\frac{\mathrm{Br}(\overline{B}_d \to \pi^+ K^-)}{\mathrm{Br}(B^+ \to \pi^+ K^0)} = 1 + r^2 - 2r\cos(\delta - \gamma) \,, \tag{6.73}$$

where δ is the strong phase difference between P and T, and $r = |T|/|P|$.

It has been argued in [21] that the combined branching ratios

$$\mathrm{Br}(B^\pm \to \pi^\pm K) \equiv \frac{1}{2}\left[\mathrm{Br}(B^+ \to \pi^+ K^0) + \mathrm{Br}(B^- \to \pi^- \overline{K^0})\right] \,, \tag{6.74}$$

$$\mathrm{Br}(B_d \to \pi^\mp K^\pm) \equiv \frac{1}{2}\left[\mathrm{Br}(B_d^0 \to \pi^- K^+) + \mathrm{Br}(\overline{B_d^0} \to \pi^+ K^-)\right] \tag{6.75}$$

may be used to constrain the CKM angle γ. The quantity of interest is the ratio

$$R = \frac{\mathrm{BR}(B_d \to \pi^\mp K^\pm)}{\mathrm{BR}(B^\pm \to \pi^\pm K)} \,, \tag{6.76}$$

which can be computed in terms of r, δ and γ to be

$$R = 1 + r^2 - 2r \cos\delta \cos\gamma \,. \tag{6.77}$$

This may be solved for the product of the cosines

$$C(R, r) = \cos\delta \cos\gamma = \frac{1}{2r}(1 - R) + \frac{r}{2} \,. \tag{6.78}$$

Since $|\cos\delta| \leq 1$ on has the inequality $|\cos\gamma| \geq |C(R, r)|$, which in general requires a knowledge of the tree-to-penguin ratio r. Unfortunately, this quantity suffers from unknown hadronic uncertainties and thus (6.78) is only of limited use.

However, it has been noted in [21] that the function $C(R, r)$ has a minimum with a value less than one. This minimum occurs at $r_0 = \sqrt{1 - R}$ and the value of C becomes

$$C(R, \sqrt{1 - R}) = \sqrt{1 - R} < 1 \,. \tag{6.79}$$

Hence for $R < 1$ we can obtain a bound on γ independent of the tree-to-penguin ratio r; this bound is

$$|\cos\gamma| \geq \sqrt{1 - R} \quad \text{or} \quad |\sin\gamma| \leq \sqrt{R} \,. \tag{6.80}$$

For a value of $R < 1$ one would exclude values of γ around $90°$, which would be complementary to other bounds on γ, obtained from a global fit, for example.

Other bounds using similar ideas have been derived and tested [22]; however, current data yield [23]

$$R = 0.92 \pm 0.15 \text{ or } R < 1.07 \,, \tag{6.81}$$

making this bound inefficient.

The discussion of the bounds does not need any information about the hadronic matrix elements. However, one may use QCD factorization to calculate the matrix elements of the effective Hamiltonian. In this way one may compute the rates of various non-leptonic charmless B decays as a function of the CKM angle γ. This has been done in [24], where a detailed phenomenological analysis was performed. The result for ratios of various rates are shown in Fig. 6.6.

As mentioned above, the situation is more complicated in the system of neutral B mesons owing to B–$\overline{B}$ oscillations. If initially a B^0 has been created, there will be oscillations in accordance with (6.25) and (6.27); hence we discuss the *time-dependent* rate into a state f common to both B^0 and $\overline{B}^0$;

$$\begin{aligned} \Gamma(B^0(t) \to f) = \frac{1}{2} e^{-\Gamma t} |A(B^0 \to f)|^2 \Big[& a(f) + b(f) e^{\Delta\Gamma t} \\ & + c(f) e^{\Delta\Gamma t} \cos(\Delta m\, t) + s(f) e^{\Delta\Gamma t} \sin(\Delta m\, t) \Big] \,, \end{aligned} \tag{6.82}$$

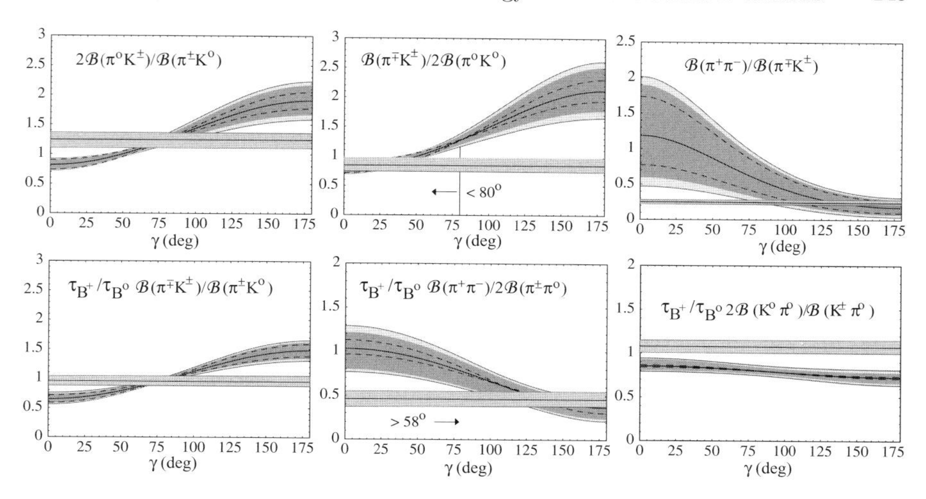

Fig. 6.6. Predictions of QCD factorization [24] for ratios of rates versus the CKM angle γ in comparison with data. The widths of the [*bands* indicate the theoretical and experimental uncertainties. The plot has been taken from [24, 25]

where we obtain, from the discussion in Sect. 6.2,

$$a(f) = \frac{1}{2}\left(1 + \left|\frac{q}{p}\overline{R}(f)\right|^2\right) + \mathrm{Re}\left(\frac{q}{p}\overline{R}(f)\right) , \tag{6.83}$$

$$b(f) = \frac{1}{2}\left(1 + \left|\frac{q}{p}\overline{R}(f)\right|^2\right) - \mathrm{Re}\left(\frac{q}{p}\overline{R}(f)\right) , \tag{6.84}$$

$$c(f) = 1 - \left|\frac{q}{p}\overline{R}(f)\right|^2 \tag{6.85}$$

$$s(f) = -2\mathrm{Im}\left(\frac{q}{p}\overline{R}(f)\right) , \tag{6.86}$$

$$\overline{R}(f) = \frac{A(\overline{B}^0 \to f)}{A(B^0 \to f)} . \tag{6.87}$$

In the same way, we obtain the following for the time-dependent decay rate into the state $\overline{f}$:

$$\Gamma(\overline{B}^0(t) \to \overline{f}) = \frac{1}{2}e^{-\Gamma t}|A(\overline{B}^0 \to \overline{f})|^2 \Big[\overline{a}(\overline{f}) + \overline{b}(\overline{f})e^{\Delta\Gamma t} + \overline{c}(\overline{f})e^{\Delta\Gamma t}\cos(\Delta m\, t) + \overline{s}(\overline{f})e^{\Delta\Gamma t}\sin(\Delta m\, t)\Big] , \tag{6.88}$$

where the quantities $\overline{a}, \ldots, \overline{s}$ are the same as $a, \ldots, s$ with the replacements

$$\frac{p}{q} \to \frac{q}{p} , \quad \overline{R}(f) \to R(\overline{f}) = \frac{A(B^0 \to \overline{f})}{A(\overline{B}^0(t) \to \overline{f})} . \tag{6.89}$$

These very general formulae simplify considerably once we take into account the fact that the lifetime difference in the system of B_d mesons is very small; this may be different for B_s, where a lifetime difference of up to 20% is possible (see Sect. 5.4). Furthermore, it has been shown in Sect. 6.2 that the ratio p/q for B mesons is simply a phase, i.e.

$$\frac{p}{q} \approx \sqrt{\frac{M_{12}}{M_{12}^*}} = e^{i\phi_M} = \begin{cases} e^{2i\beta} \text{ for } B_b \\ e^{2i\delta\gamma} \text{ for } B_s \end{cases} \tag{6.90}$$

where ϕ_M is the B–$\overline{B}$ mixing phase discussed in Sect. 6.2.

A special role is played by decay modes of neutral B mesons into CP eigenstates f_{CP}. In this case we have

$$f_{CP} = \eta \overline{f}_{CP} \, , \quad \eta = \pm 1 \, , \tag{6.91}$$

where η is the CP quantum number of the final state f_{CP}.

Taking into account the simplifications mentioned above, we obtain

$$\Gamma(B^0(t) \to f_{CP}) = \frac{1}{2} e^{-\Gamma t} |A(B^0 \to f_{CP})|^2 \times \left[d(f_{CP}) + c(f_{CP}) \cos(\Delta m\, t) + s(f_{CP}) \sin(\Delta m\, t) \right] \tag{6.92}$$

where

$$d(f_{CP}) = \frac{1}{2} \left(1 + \left| \frac{A(\overline{B^0} \to f_{CP})}{A(B^0 \to f_{CP})} \right|^2 \right) . \tag{6.93}$$

Using also (6.88), we can define a *time-dependent* CP asymmetry $\mathcal{A}_{CP}(t)$ as

$$\begin{aligned} \mathcal{A}_{CP}(t) &= \frac{\Gamma(B^0(t) \to f_{CP}) - \Gamma(\overline{B}^0(t) \to f_{CP})}{\Gamma(B^0(t) \to f_{CP}) + \Gamma(\overline{B}^0(t) \to f_{CP})} \\ &= \hat{C}(f_{CP}) \cos(\Delta m\, t) + \hat{S}(f_{CP}) \sin(\Delta m\, t) \, , \end{aligned} \tag{6.94}$$

where

$$\hat{C}(f_{CP}) = \frac{\Gamma(B^0 \to f_{CP}) - \Gamma(\overline{B}^0 \to f_{CP})}{\Gamma(B^0 \to f_{CP}) + \Gamma(\overline{B}^0 \to f_{CP})} \, , \tag{6.95}$$

$$\hat{S}(f_{CP}) = -\frac{2}{1 + |\overline{R}(f_{CP})|^2} \mathrm{Im} \left[e^{i\phi_M} \overline{R}(f_{CP}) \right] . \tag{6.96}$$

Note that the first term is simply the direct CP asymmetry (6.70) for the neutral B mesons, while the second term is due to mixing and is called the *mixing-induced* CP asymmetry. A non-vanishing contribution to the first term can only from the existence of a weak *and* a strong phase difference between different contributions to the rate, while the second term appears because the time evolution itself generates a "strong" phase difference $\exp(i\Delta m\, t)$ between the two neutral B meson states, which – together with the weak phases – leads to a CP asymmetry.

As stated above, the B mesons have numerous decay channels and it is beyond the scope of this book to consider all the possible channels. We refer the interested reader to excellent reviews such as [26] for the various strategies to extract CKM angles from various decay modes.

CP violation studies mainly use exclusive non-leptonic decays, for which no reliable theoretical description, based for example on an effective-field-theory picture derived from QCD, exists at present. The strategies discussed in [26] are based mainly on flavour symmetries sometimes combined with plausible dynamical assumptions, and, since they are not an application of effective-field-theory methods, we shall not consider them here.

For this reason we shall restrict ourselves to classifying the various possible modes by their underlying quark transitions. Thus we have to study the effective Hamiltonian for the weak decays discussed in Sect. 5.1 and collect the terms with weak phases. At tree level, we look at the contributions given in (5.9)–(5.16) and identify the terms that carry weak phases, where we shall use the standard convention for the phases introduced in (3.11). In this convention all CKM elements appearing in (5.9)–(5.16) are, to leading order in the Wolfenstein parametrization, real except for the matrix element V_{ub}.

If we were to assume for the moment that the tree-level diagram were the only contributions to a decay, we would only have a single amplitude and hence no direct CP violation could occur. This is manifest, since we then would have

$$|A(B^0 \to f_{CP})| = |A(\overline{B}^0 \to f_{CP})| \to |\overline{R}(f_{CP})| = 1 \; . \tag{6.97}$$

However, the time-dependent CP asymmetry may still be non-vanishing due to the weak phase in the mixing; for such a mixing-induced CP asymmetry we obtain from (6.94)

$$\mathcal{A}_{CP}(t) = -\mathrm{Im}\left[e^{(i\phi_M + 2\phi_{f_{CP}})}\right] \sin(\Delta m\, t) \; , \tag{6.98}$$

where $\phi_{f_{CP}}$ is the weak decay phase of the single contribution.

In this way, we can easily classify the possible decay modes. All modes which have a vanishing decay phase are sensitive to the mixing phase, which is 2β for B_d decays and $2\delta\gamma$ for B_s decays. Thus, for B_d decays where the amplitude carries no weak phase (i.e. the processes due to the quark transition $b \to c\bar{q}q'$ with $q = u, c$ and $q' = s, d$), we obtain

$$\mathcal{A}_{CP}^{b \to c\bar{q}q'}(t) = -\sin(2\beta)\sin(\Delta m\, t) \; . \tag{6.99}$$

Similarly, for B_d decays with a decay amplitude proportional to V_{ub} (i.e. the processes due to the quark transition $b \to u\bar{q}q'$ with $q = u, c$ and $q' = s, d$) we find

$$\mathcal{A}_{CP}^{b\to u\bar{q}q'}(t) = -\sin(2\beta + 2\gamma)\sin(\Delta m\, t) = \sin 2\alpha \sin(\Delta m\, t) \;, \tag{6.100}$$

where we have used the triangle relation $\alpha + \beta + \gamma = \pi$.

However, including QCD corrections and taking into account the fact that in general two or more amplitudes, which can have different weak phases, can contribute complicates the situation. In particular, a direct CP violation (i.e. a term proportional to $\cos(\Delta m\, t)$ will appear. In order to study this in detail we have to consider individual decay modes. However, discussing all interesting modes is beyond the scope of this book, and we shall pick only two examples.

The mode which was considered most intensively in the first years of the B factories was the mode $B^0 \to J/\Psi K_s$, originating from the quark transition $b \to c\bar{c}s$. Here the combination of quarks in the final state is not a CP eigentstate; however, as discussed in Sect. 5.3, the state K_s is a coherent superposition of the quark states $\bar{s}d$ and $\bar{d}s$. If we now consider the CP-conjugate process, we have the quark transition $\bar{b} \to \bar{c}c\bar{s}$, which has the same matrix element with the K_s state.[1]

As discussed above the combination of CKM matrix elements appearing in the decay $B^0 \to J/\Psi K_s$ is real, and hence we expect

$$\overline{R}(J/\Psi K_s) = 1 \tag{6.101}$$

and

$$\mathcal{A}_{CP}^{B^0\to J/\Psi K_s}(t) = -\sin 2\beta \sin(\Delta m\, t) \;. \tag{6.102}$$

Once the full effective Hamiltonian, as discussed in Sect. 5.1, is taken into account, we have to calculate radiative corrections due to gluon exchange. Considering the full $b \to s$ effective Hamiltonian, we have to include contributions from radiative corrections which induce penguin contributions. Looking at the CKM factors of the penguin contributions we find that the leading contribution is that from the charm quark, and has the CKM factor $V_{cb}V_{cs}^*$, which is the same CKM factor as that of the contributions from tree level. Thus this contribution carries the same weak phase and hence will not lead to any direct CP violation. The remaining penguin contributions in fact carry a different weak phase (such as the quark transition $b \to u\bar{u}s$, which carries the phase $\exp(-i\gamma)$ in the usual convention), but these are strongly CKM suppressed. Thus we have in the Standard Model

$$\overline{R}(J/\Psi K_s) = 1 + \mathcal{O}(\lambda^2) \;, \tag{6.103}$$

where λ is the Wolfenstein parameter of the CKM matrix introduced in (3.13) and (3.14). As a consequence, we expect that (6.102) will hold at the level of a few per cent, and thus this so-called gold-plated mode allows a clean determination of the CKM angle β [27].

[1] We ignore here the tiny effect of CP violation in the kaon system.

In fact, the current data from the B factories already allow a significant measurement of the CKM phase β. The current average is [28]

$$\sin 2\beta = 0.736 \pm 0.049 . \tag{6.104}$$

which already has an accuracy below 10%, meaning that CP violation in the B system is clearly established. In fact, this result may be used to constrain possible new-physics effects [29].

The same discussion can be performed for the decays $B^0 \to \pi^+\pi^-$ and $B^0 \to \pi^0\pi^0$, which are mediated by the same quark transition $b \to u\bar{u}d$. At tree level we would expect

$$\overline{R}(\pi^+\pi^-) = e^{2i\gamma} = \overline{R}(\pi^0\pi^0) , \tag{6.105}$$

which, as discussed above, leads to the naive result

$$\mathcal{A}_{CP}^{B^0\to\pi^+\pi^-}(t) = \mathcal{A}_{CP}^{B^0\to\pi^0\pi^0}(t) = \sin 2\alpha \sin(\Delta m\, t) . \tag{6.106}$$

However, in this case the situation becomes more complicated once the full effective Hamiltonian is taken into account. Here we have to consider the full effective interaction for a $b \to u$ transition, which contains penguin contributions that could lead to a sizeable direct CP violation. From the point of view of isospin, the situation is very similar to that in the decays $K \to \pi\pi$: there are two contributions with $\Delta I = 1/2$ and $\Delta I = 3/2$, which can interfere, having in general different strong phases. The $\Delta I = 3/2$ contribution comes purely from the tree diagram and thus carries the weak phase $e^{-i\gamma}$, while the penguins induce only $\Delta I = 1/2$ contributions and carry, together with the $\Delta I = 1/2$ piece of the tree contribution, a different weak phase. However, here there is no $\Delta I = 1/2$ rule at work (see (5.126)), both amplitudes are expected to be of the same order. The leading contribution to the penguins will be the one from the operator $b \to c\bar{c}d$, which is of the same order in the Wolfenstein parameter λ as the tree-level contribution, and thus no suppression of the penguin contribution is expected. Thus we have

$$|\overline{R}(\pi^+\pi^-)| \neq 1 , \quad |\overline{R}(\pi^0\pi^0) \neq 1 . \tag{6.107}$$

In order to obtain a quantitative statement about this decay, we have to get some information on the size and the (strong) phase of the penguin amplitudes relative to the tree amplitudes. One way to do this is by using isospin relations, which need full information about all $B \to \pi\pi$ decays; the details of this method can be found in [30].

Alternatively, one can use QCD factorization to calculate the relative sizes and phases of the penguin contribution. This analysis has been performed in [24], where the time-dependent CP asymmetry for $B_d \to \pi^+\pi^-$ has been computed. Owing to the penguin contribution, $\hat{S}(\pi^+\pi^-)$ deviates from $\sin 2\alpha$, and the deviation is a function of the weak phase γ. One may compute the

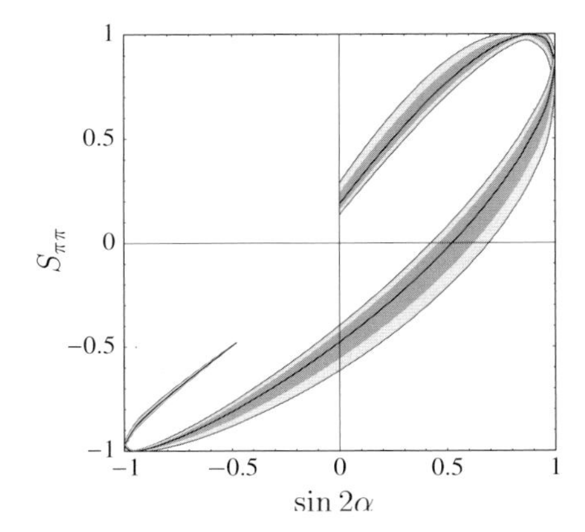

Fig. 6.7. Allowed region in the $S_{\pi\pi}$–$\sin 2\alpha$ plane as predicted from QCD factorization. The width of the *band* indicates the theoretical uncertainty. Plot taken from [24]

relation between $\hat{S}(\pi^+\pi^-)$ and $\sin(2\alpha)$ in QCD factorization, and the result is shown in Fig. 6.7.

Once time-dependent CP aymmetries are measured in the B_s system, a more precise determination of the CKM angles will become possible. This is mainly due to the fact that the mixing phase in the B_s system is very small, and a stringent test of the Standard Model will be provided by the measurement of this mixing phase in a decay such as $B_s \to J/\Psi\phi$. Furthermore, the B_s system will also open the road to clean determinations of the CKM angle γ; however, a discssion of this matter is beyond the scope of this book and we have to refer the reader to excellent reviews such as [26].

References

1. I. I. Y. Bigi and A. I. Sanda, *CP Violation*, Cambridge Monographs on Particle Physics, Nuclear Physics and Cosmology, No. 9 (Cambridge University Press, Cambridge, 2000).
2. O. Nachtmann, *Elementary Particle Physics*, Springer Texts and Monographs in Physics (Springer, Berlin, Heidelberg, 1989).
3. T. Inami and C. S. Lim, Prog. Theor. Phys. **65**, 297 (1981); erratum, Prog. Theor. Phys. **65**, 1772 (1981).
4. F. J. Gilman and M. B. Wise, Phys. Rev. D **27**, 1128 (1983).
5. G. Buchalla, A. J. Buras and M. E. Lautenbacher, Rev. Mod. Phys. **68**, 1125 (1996) [arXiv:hep-ph/9512380].
6. S. Herrlich and U. Nierste, Nucl. Phys. B **476**, 27 (1996) [arXiv:hep-ph/9604330].
7. D. Becirevic, plenary talk at the 21st International Symposium on Lattice Field Theory (LATTICE 2003), Tsukuba, Ibaraki, Japan, 15–19 Jul. 2003.

8. M. Beneke, G. Buchalla, C. Greub, A. Lenz and U. Nierste, Phys. Lett. B **459**, 631 (1999) [arXiv:hep-ph/9808385].
9. S. M. Ryan and A. S. Kronfeld, Nucl. Phys. Proc. Suppl. **119**, 622 (2003) [arXiv:hep-lat/0209083].
10. K. Hagiwara et al. [Particle Data Group], Phys. Rev. D **66**, 010001 (2002).
11. I. I. Y. Bigi and A. I. Sanda,
12. H. Georgi, Phys. Lett. B **297**, 353 (1992).
13. T. Ohl, G. Ricciardi, E. H. Simmons, Nucl. Phys. B **403**, 605 (1993) [arXiv:hep-ph/9301212].
14. J. Christensen, J. Cronin, V. Fitch and R. Turlay, Phys. Rev. Lett. **13**, 138 (1964); Phys. Rev. **140** B74 (1965).
15. L. Wolfenstein, Phys. Rev. Lett. **13**, 562 (1964).
16. A. J. Buras and M. Jamin, JHEP **0401**, 048 (2004) [arXiv:hep-ph/0306217].
17. A. J. Buras, M. Jamin and M. E. Lautenbacher, Phys. Lett. B **389**, 749 (1996) [arXiv:hep-ph/9608365].
18. M. Ciuchini, E. Franco, L. Giusti, V. Lubicz and G. Martinelli, contribution to Conference on Kaon Physics (K 99), Chicago, 21-26 Jun. 1999 [arXiv:hep-ph/9910237].
19. S. Bertolini, J. O. Eeg, M. Fabbrichesi and E. I. Lashin, Nucl. Phys. B **514**, 93 (1998) [arXiv:hep-ph/9706260].
20. S. Bertolini, AIP Confonference Proceedings, No. 618, p. 79 (American Institute of Physics, New York, 2002) [arXiv:hep-ph/0201218].
21. R. Fleischer and T. Mannel, Phys. Rev. D **57**, 2752 (1998) [arXiv:hep-ph/9704423].
22. M. Neubert and J. L. Rosner, Phys. Lett. B **441**, 403 (1998) [arXiv:hep-ph/9808493].
23. Heavy Flavor Averaging Group, http://www.slac.stanford.edu/xorg/hfag/
24. M. Beneke, G. Buchalla, M. Neubert and C. T. Sachrajda, Nucl. Phys. B **606**, 245 (2001) [arXiv:hep-ph/0104110].
25. S. Stone, plenary talk at the International Europhysics Conference on High-Energy Physics (HEP 2003), Aachen, Germany, 17–23 Jul. 2003 [arXiv:hep-ph/0310153].
26. R. Fleischer, Phys. Rep. **370**, 537 (2002) [arXiv:hep-ph/0207108].
27. I. I. Y. Bigi and A. I. Sanda,
28. T. E. Browder, plenary talk at the 21st International Symposium on Lepton and Photon Interactions at High Energies (LP 03), Batavia, Illinois, 11–16 Aug 2003 [arXiv:hep-ex/0312024].
29. R. Fleischer and T. Mannel, Phys. Lett. B **506**, 311 (2001) [arXiv:hep-ph/0101276].
30. M. Gronau and D. London, Phys. Rev. Lett. **65**, 3381 (1990).

7 Beyond the Standard Model

As of today, the Standard Model has passed all tests and no significant hint of any physics beyond this model has been found. On the other hand, being the most general renormalizable theory (if the minimal Higgs sector is included) with the desired particle spectrum and the observed interactions, it is mainly parametrizing our ignorance about physics at higher scales. This parametrization requires 27 parameters, of which only three (the gauge couplings) are generically related to its gauge theory structure. All other parameters originate from the symmetry-breaking sector. We have as a possible set of parameters the electroweak vacuum-expectation value, the Higgs mass, six quark masses and six lepton masses, three mixing angles in the quark sector and another three for the leptons, and four CP-violating phases, of which three originate from the lepton sector, assuming Majorana neutrinos.

It has to be considered a great success to be able to describe all known phenomena in terms of these parameters, but it is also, on the other hand, unsatisfactory not to be able to compute, for example, the ratio of the electron mass to the muon mass. However, the observed structure of the parameters may lead to interesting clues about what is behind the Standard Model. Some open questions of this kind are

- Why are the off-diagonal matrix elements of the CKM matrix so small? Why is the CKM matrix in this sense hierarchical, while its leptonic counterpart, the MNS matrix, is (as far as we can tell today) not?
- Why are the quark masses (except for the top quark) so small compared with the electroweak vacuum expectation value? Does the top quark (or, more generally, the third generation) play a special role?
- Why do we observe three families? Is there a symmetry behind this?
- Why is CP violation so small? Why is CP violation in flavour-neutral processes (such as electric dipole moments of particles) not observed?

It is the current hope that we shall be able to gain some insight into these questions by testing the flavour structure of the Standard Model in some detail. While the experiments at LEP and SLC have tested the gauge structure of the Standard Model with extreme precision, it will be hard to obtain similarly accurate information about the parameters in the flavour sector, since severe limitations due to hadronic uncertainties exist here, at least at present.

Theoretical ideas about physics beyond the Standard Model have been strongly influenced by the success of gauge theories. For example, grand unified theories (GUTs) have been considered as a natural extension of the Standard Model, but GUTs still simply triplicate the particle content to take into account the three families. Except for the fact that leptons and quarks are members of the same multiplet of the group used for grand unification, and thus their masses have to be equal at the GUT scale, these theories do not provide any ansatz that might answer all or at least some of the above questions.

Supersymmetric theories have been widely used to parametrize possible physics beyond the Standard Model, for example in the analysis of high-energy-physics data. As far as the symmetry-breaking sector and the mixing of flavours are concerned, the situation in a supersymmetric theory becomes much worse than in the Standard Model. Aside from the fact that the three generations are introduced by hand just as in the Standard Model, the Higgs sector needs to be extended in order to comply with supersymmetry. Furthermore, the duplication of the particle spectrum (i.e. the introduction of the supersymmetric partners) yields many more sources of flavour mixing and CP violation, and a serious fine tuning (or some other special assumption) is needed for the theory to be consistent with data. In particular, the observed small CP violation, appearing only in the charged-current sector, cannot be introduced into a supersymmetric theory in a natural way. On the basis of our current knowledge, it is fair to say that supersymmetry clearly has a flavour problem.

There are many more ideas about how to extend the Standard Model, but generically these ideas run into problems when it comes to flavour. In particular, most extensions yield additional sources of CP violation, in most cases also in the flavour-neutral sector. In all those cases some fine tuning is needed to adjust the couplings in such a way that experimental constraints are met.

Unfortunately, as of today, we do not have any theory of flavour nor do we have an efficient parametrization of deviations from the flavour sector of the Standard Model, such as the Peskin–Takeuchi parameters [1], which have been used in the gauge sector. In the next section we shall discuss, in an effective-theory language, why this is so much more difficult in the flavour sector.

It is tempting to use symmetry arguments to explain the flavour structure of the Standard Model. Such a family symmetry (or horizontal symmetry) is restricted by the observed flavour structure, and we discuss some general properties in Sect. 7.2.

7.1 The Standard Model as an Effective Field Theory

One generic way to discuss effects of physics beyond the Standard Model is to look at the Standard Model as the renormalizable piece of an effective theory, obtained from integrating out the new physics happening at some large scale Λ. Any effect of physics at the large scale Λ is at scales of the order of the electroweak vacuum expectation value encoded in higher-dimensional operators.

We can turn this argument around and write all possible operators of higher dimension with unknown coupling constants, thereby parametrizing any scenario of new physics. Postulating invariance under the electroweak symmetry group $SU(3)_{QCD} \times SU(2)_W \times U(1)_Y$ we find that the lowest-dimensional operators are those of dimension six, which are thus suppressed by two powers of the large scale Λ.

Thus we have in general

$$\mathcal{L} = \mathcal{L}_{SM} + \frac{1}{\Lambda^2} \sum_i g_i \mathcal{O}_i \; , \tag{7.1}$$

where the $\mathcal{O}_i$ are all possible dimension-six operators compatible with the Standard Model symmetry, i.e. $SU(3)_{QCD} \times SU(2)_W \times U(1)_Y$.

The possible dimension-six operators were considered some time ago [2]; they fall into three classes. There is one class which does not contain any quark or lepton field, that is, these operators consist only of gauge and Higgs fields. Clearly, these operators are irrelevant with respect to flavour physics, but they have been used to parametrize new-physics effects in the gauge sector [3, 4]. In fact, one may reformulate the analysis of Peskin and Takeuchi [1] in terms of this class of dimension-six operators [4]. Clearly these operators cannot mix under renormalization with any operator having a non-trivial flavour structure.

The operators having a non-trivial flavour structure can be divided into two classes. The first class are four-fermion operators, the second class consists of operators with two quark or lepton fields, where the remaining fields needed to obtain a dimension-six operator are gauge and Higgs fields. After using the equations of motion, we have the operators

$$O_{LL}^{(1)} = \overline{Q}_A \not{L} G_{AB}^{(1)} Q_B \; , \tag{7.2}$$

$$O_{LL}^{(2)} = \overline{Q}_A \not{L}_3 G_{AB}^{(2)} Q_B \; , \tag{7.3}$$

$$O_{RR}^{(1)} = \overline{q}_A \not{R} F_{AB}^{(1)} q_B \; , \tag{7.4}$$

$$O_{RR}^{(2)} = \overline{q}_A \left\{ \tau_3, \not{R} \right\} F_{AB}^{(2)} q_B \; , \tag{7.5}$$

$$O_{RR}^{(3)} = i\overline{q}_A \left[\tau_3, \not{R} \right] F_{AB}^{(3)} q_B \; , \tag{7.6}$$

$$O_{RR}^{(4)} = \overline{q}_A \tau_3 \not{R} \tau_3 F_{AB}^{(4)} q_B \; , \tag{7.7}$$

$$O_{LR}^{(1)} = \overline{Q}_A H H^\dagger H \hat{K}_{AB}^{(1)} q_B + \text{h.c.} , \tag{7.8}$$

$$O_{LR}^{(2)} = \overline{Q}_A \left(\sigma_{\mu\nu} B^{\mu\nu}\right) H \hat{K}_{AB}^{(2)} q_B + \text{h.c.} , \tag{7.9}$$

$$O_{LR}^{(3)} = \overline{Q}_A \left(\sigma_{\mu\nu} W^{\mu\nu}\right) H \hat{K}_{AB}^{(3)} q_B + \text{h.c.} , \tag{7.10}$$

$$O_{LR}^{(4)} = \overline{Q}_A \left(i D_\mu H\right) i D^\mu \hat{K}_{AB}^{(4)} q_B + \text{h.c.} , \tag{7.11}$$

with the definitions

$$L^\mu = H \left(i D^\mu H\right)^\dagger + \left(i D^\mu H\right) H^\dagger , \tag{7.12}$$

$$L_3^\mu = H \tau_3 \left(i D^\mu H\right)^\dagger + \left(i D^\mu H\right) \tau_3 H^\dagger , \tag{7.13}$$

$$R^\mu = H^\dagger \left(i D^\mu H\right) + \left(i D^\mu H\right)^\dagger H , \tag{7.14}$$

and

$$\hat{K}_{AB}^{(i)} = K_{AB}^{(i)} + \tau_3 K_{AB}^{(i)\prime} . \tag{7.15}$$

Note that the operators involving explicit factors of τ_3 violate the custodial $SU(2)$ symmetry, while the other operators conserve that symmetry.

The flavour structure is encoded in the coupling matrices $G_{AB}^{(i)}$, $F_{AB}^{(i)}$, $K_{AB}^{(i)}$ and $K_{AB}^{(i)\prime}$, leaving an enormous number of parameters. To reduce the number of parameters, one has to make restrictive assumptions, for example by using a specific model, from which these matrices can be computed. We shall not go into any detail about this, and instead refer the reader to recent attempts along these lines [5, 6].

Even more parameters are present once the four-fermion operators are included, since there are many possible Dirac structures, as well as two possible colour structures [2]. In order to give a flavour of the variety, we discuss first the four-quark operators corresponding to charged currents for $\Delta B = 1$.

A general charged-current $\Delta B = \pm 1$ interaction is given by

$$\mathcal{C}_{q;\alpha;\lambda\lambda'\sigma\sigma'}^{(C)kl} = (\bar{b}_\lambda \Gamma_\alpha C U_{\lambda'}^k)(\bar{U}_\sigma^l \Gamma_\alpha C q_{\sigma'}) , \tag{7.16}$$

where U_i labels the up quarks and q is either s or d. The subscript α refers to the Dirac structure of the operator, for which we have the usual five possibilities:

$$\Gamma_\alpha \otimes \Gamma_\alpha = \begin{cases} 1 \otimes 1 \\ \gamma_\mu \otimes \gamma^\mu \\ \sigma_{\mu\nu} \otimes \sigma^{\mu\nu} \\ \gamma_\mu \gamma_5 \otimes \gamma_5 \gamma^\mu \\ \gamma_5 \otimes \gamma_5 \end{cases} . \tag{7.17}$$

The superscript $C = 1, 8$ refers to the colour structure, for which we have

$$C \otimes C = \begin{cases} 1 \otimes 1 & \text{for } C = 1 \\ T^a \otimes T^a & \text{for } C = 8 \end{cases} , \tag{7.18}$$

the indices $\lambda, \lambda', \sigma, \sigma' = \text{L, R}$ label the left and right handed helicity components of the quarks, and k, l are the flavour indices referring to the up-type quarks.

Similarly, a general $\Delta B = \pm 1$ neutral-current operator contains only down-type quarks and can be written as

$$\mathcal{N}^{(C)klm}_{\alpha;\lambda\lambda'\sigma\sigma'} = (\bar{b}_\lambda \Gamma_\alpha C D^k_{\lambda'})(\bar{D}^l_\sigma \Gamma_\alpha C D^m_{\sigma'}) , \tag{7.19}$$

where the notation concerning the indices is the same as for the charged-current interaction.

Thus the most general effective Hamiltonian for $\Delta B = 1$ will be a linear combination of these operators with coefficients depending on the new physics. Clearly, there are too many parameters for a generic description through this most general effective Hamiltonian.

The number of possible operators can in principle be reduced by using a Fierz rearrangement of the quark fields, which will relate some of the operators to each other. We shall not discuss this here, since there will be still too large a number of operators to make this general approach useful.

7.2 Flavour in Models Beyond the Standard Model

The fact that three families exist in which the particles have identical quantum numbers with respect to $SU(3)_{QCD} \times SU(2)_W \times U(1)_Y$ suggests strongly that a symmetry, called a family symmetry or horizontal symmetry, lies behind this triplication of the observed particle spectrum. However, it is fair to say that as of now there is neither a generally accepted nor a predictive framework for flavour.

In this section we shall discuss the general properties of a possible flavour symmetry. It will become clear what the problems are, and how a possible scenario which explains the hierarchical structure of the masses and the CKM matrix could look. These ideas are in fact quite old and date back to the classic paper by Froggatt and Nielsen [7]; they recently have been discussed in a more general framework in [8, 9].

A horizontal symmetry group F has to satisfy certain constraints which can be discussed very generically. The general assumption is that such a symmetry gives the observed structure to the quark mass matrices. We shall first prove two well-known facts about the symmetry group F:

- The symmetry F cannot be exact, i.e. it has to be broken.
- The simple scalar sector of the Standard Model has to be extended.

In order to discuss these issues we shall first define the action of H on the fields introduced in Chap. 2. Since the horizontal symmetry is assumed to commute with the Standard Model gauge symmetry $SU(3)_{QCD} \times SU(2)_W \times U(1)_Y$, we have the following for the quark fields[1] in the notation of Chap. 2:

[1] We discuss these issues for the quarks, the same can be done for the leptons.

$$Q_A \to (F_L)_{AB} Q_B\,, \quad (P_+ q)_A = (F_u)_{AB}(P_+ q)_B\,, \quad P_- q_A = (F_d)_{AB} P_- q_B\,, \tag{7.20}$$

where $P_\pm = (1 \pm \tau_3)/2$ projects out up- and down-type quarks. For a compact notation we shall suppress the family indices in the following, i.e. we write

$$Q \to F_L Q\,, \quad (P_+ q) = F_u (P_+ q)\,, \quad (P_- q) = F_d (P_- q)\,. \tag{7.21}$$

Note that F_L, F_u and F_d are three-dimensional unitary representations of the horizontal symmetry F.

As has been discussed in Chap. 2, we do not need to consider the gauge fields to understand the flavour structure of the Standard Model. However, we have to discuss the Higgs field. Since there is only a single Higgs doublet in the Standard Model, it has to transform under a one-dimensional representation which is similar to a $U(1)_Y$ transformation, i.e.

$$H \to H \exp(i\phi\tau_3)\,, \tag{7.22}$$

where ϕ is a phase.

The only couplings of interest for the discussion of the horizontal symmetry and its effect on masses and mixings are the Yukawa couplings. We write these couplings using the above compact notation as

$$\mathcal{L}_{\mathrm{Yuk}} = \frac{1}{v}\left(\bar{Q}\mathcal{M}_u H(P_+ q) + \bar{Q}\mathcal{M}_d H(P_- q)\right)\,, \tag{7.23}$$

where $\mathcal{M}_{u/d}$ are the up/down quark mass matrices.

Performing now a symmetry transformation of the horizontal symmetry F and requiring that $\mathcal{L}_{\mathrm{Yuk}}$ is invariant under F, we find that

$$\mathcal{M}_u = F_L^\dagger \mathcal{M}_u F_u\, e^{i\phi}\,, \quad \mathcal{M}_d = F_L^\dagger \mathcal{M}_d F_d\, e^{-i\phi}\,, \tag{7.24}$$

from which we obtain

$$\left[\mathcal{M}_u \mathcal{M}_u^\dagger\,,\, F_L\right] = 0\,, \quad \left[\mathcal{M}_d \mathcal{M}_d^\dagger\,,\, F_L\right] = 0\,, \tag{7.25}$$

$$\left[\mathcal{M}_u^\dagger \mathcal{M}_u\,,\, F_u\right] = 0\,, \quad \left[\mathcal{M}_d^\dagger \mathcal{M}_d\,,\, F_d\right] = 0\,. \tag{7.26}$$

From the relations (7.25), we may draw a few interesting conclusions. Assuming that the representations of the quark fields with respect to F are irreducible, we find that both quark mass matrices have to be proportional to the unit matrix, i.e. we obtain degenerate quark masses. Alternatively, the representations can be reducible, in which case F_L, F_u and F_d can be diagonal, and the CKM matrix becomes trivial. Since there are both non-degenerate quark masses and non-trivial mixing, the symmetry F has to be broken.

The second observation, namely that F cannot be spontaneously broken by the vacuum expectation value of a single Higgs field (i.e. the Standard

Model case), follows from the fact that the transformation of the Higgs field under F is the same as a hypercharge transformation. This implies that the phase ϕ can always be removed by a compensating hypercharge transformation, in which case the Higgs field can be regarded as invariant under F. Since an invariant field can never break a symmetry by acquiring a vacuum expectation value, we may apply the above steps again to conclude that we have to have either degenerate quarks or vanishing mixing. In turn, the scalar sector needs to be extended.

Given the large splitting of the quark masses any non-abelian horizontal symmetry will effectively become a abelian symmetry. If a non-abelian symmetry is introduced, all up-type quarks are in the same mulitiplet and different masses can be generated only by a large breaking of this symmetry. In order to study the mixing between families, it will be sufficient to consider an abelian symmetry.

It is well known that introducing more than one Higgs field carries the danger of large flavour-changing neutral currents. The simplest extension is a two-Higgs-doublet model, of which only two types are safe with respect to flavour changing neutral currents. We discuss the type of two-Higgs-doublet models which is also relevant for supersymmetric theories. The Lagrangian for the Yukawa terms reads

$$\mathcal{L}_{\rm Yuk}^{\rm IIHDM} = \frac{1}{v} \left(\bar{Q} \mathcal{M}_u H_u (P_+ q) + \bar{Q} \mathcal{M}_d H_d (P_- q) \right) , \tag{7.27}$$

where we have introduced two Higgs fields H_u and H_d, giving masses to the up and down quarks. The two Higgs fields are again in a one-dimensional representation of the horizontal symmetry and transform as

$$H_u \to H_u \exp(i\phi_u \tau_3) , \quad H_d \to H_d \exp(i\phi_d \tau_3) . \tag{7.28}$$

The Yukawa interaction (7.27) still has a symmetry on top of the $U(1)_Y$ that has been used above, which is a transformation of the form

$$H_d \to H_d \exp(i\psi \tau_3) , \quad d \to d \exp(i\psi) , \tag{7.29}$$

while all other fields remain unchanged. Thus, by the same argument as for the single Higgs doublet of the Standard Model, we can compensate the phase of the horizontal-symmetry transformation by adjusting the phase ψ in (7.29). Thus the two Higgs fields H_u and H_d again cannot break the horizontal symmetry.

One way to avoid these "no-go" statements concerning the spontaneous breaking of the horizontal symmetry is to introduce non-renormalizable terms suppressed by a large scale Λ. This is natural since it introduces the mixing between families as a power-suppressed contribution, explaining the smallness of this effect. We wish only to discuss these models schematically, so we shall consider a simplified model, where the horizontal symmetry is simply a $U(1)_H$ phase transformation. We shall define the two Higgs doublets to be in the trivial representation of the horizontal symmetry, i.e. $\phi_u = 0 = \phi_d$. We introduce an additional scalar field S, which carries one unit of charge under the horizontal symmetry

$$S \rightarrow S \exp(-i\phi) , \tag{7.30}$$

while the quarks transform as

$$F_L = \exp(iT_L\phi) , \quad F_u = \exp(iT_u\phi) , \quad F_d = \exp(iT_d\phi) . \tag{7.31}$$

The charges T are 3×3 matrices that can be chosen to be diagonal:

$$T_L = \begin{pmatrix} t_L^{(1)} & 0 & 0 \\ 0 & t_L^{(2)} & 0 \\ 0 & 0 & t_L^{(3)} \end{pmatrix} , \quad T_u = \begin{pmatrix} t_u^{(1)} & 0 & 0 \\ 0 & t_u^{(2)} & 0 \\ 0 & 0 & t_u^{(3)} \end{pmatrix} , \quad T_d = \begin{pmatrix} t_d^{(1)} & 0 & 0 \\ 0 & t_d^{(2)} & 0 \\ 0 & 0 & t_d^{(3)} \end{pmatrix} , \tag{7.32}$$

where the diagonal entries are assumed to be integer numbers.

A $U(1)_H$-invariant, non-renormalizable Yukawa interaction term can now be written as

$$\begin{aligned} \mathcal{L}_{\text{Yuk}}^{\text{nr}} = \Bigg(& \frac{1}{\Lambda^{n_{AB}}} \bar{Q}_A \lambda_{u,AB} S^{n_{AB}} H_u (P_+ q)_B \\ & + \frac{1}{\Lambda^{m_{AB}}} \bar{Q}_A \lambda_{d,AB} S^{m_{AB}} H_d (P_- q)_B \Bigg) , \end{aligned} \tag{7.33}$$

where we have re-inserted the family indices A and B, and Λ is a scale of new physics which induces the nonrenomalizable terms (7.33). The powers of the field S are determined from the requirement that the Lagrangian $\mathcal{L}_{\text{Yuk}}^{\text{nr}}$ is invariant under $U(1)_H$, which yields the relations

$$n_{AB} = t_L^{(A)} - t_u^{(B)} , \quad m_{AB} = t_L^{(A)} - t_d^{(B)} . \tag{7.34}$$

Spontaneous breaking of the horizontal symmetry means that S acquires a vacuum expectation value, which we write as

$$\langle S \rangle = \epsilon \Lambda , \tag{7.35}$$

introducing a small quantity ϵ.

Hierarchical mass matrices and mixing can now be introduced by a suitable choice of the charges for the different generations, and the hierarchy is determined by the small parameter epsilon. Assuming for purposes of illustration $\lambda_{u,AB} = 1$ and $\lambda_{d,AB} = 1$ for all values of A and B, we find a mass term in the Lagrangian of the form

$$\mathcal{L}_{Mass} = v \left(\bar{Q}_A \epsilon^{n_{AB}} (P_+ q)_B + \bar{Q}_A \epsilon^{m_{AB}} (P_- q)_B \right) , \tag{7.36}$$

where we have used a simplified picture in which $\langle H_u \rangle = \langle H_d \rangle = v$.

In this simple picture, the mixing angles are related to the charge differences between the left-handed components of the families; we obtain, as order-of-magnitude relations,

$$|V_{us}| \sim \epsilon^{t_L^{(2)}-t_L^{(1)}} , \quad |V_{cb}| \sim \epsilon^{t_L^{(3)}-t_L^{(2)}} , \quad |V_{ub}| \sim \epsilon^{t_L^{(3)}-t_L^{(1)}} , \tag{7.37}$$

which implies a structure corresponding to the Wolfenstein parametrization. In particular, we obtain

$$|V_{ub}| \sim |V_{cb}||V_{us}| . \tag{7.38}$$

Similarly, the mass ratios depend on the charges of the right-handed particles; we find, as order-of-magnitude estimates,

$$\frac{m_d^{(A)}}{m_d^{(B)}} \sim \epsilon^{t_L^{(A)}-t_L^{(B)}+t_d^{(A)}-t_d^{(B)}} , \quad \frac{m_u^{(A)}}{m_u^{(B)}} \sim \epsilon^{t_L^{(A)}-t_L^{(B)}+t_u^{(A)}-t_u^{(B)}} . \tag{7.39}$$

We shall not go into any quantitative details here, since for a quantitatively satisfactory model this simple toy model needs to be extended [8, 9]. However, we may relate this model to the simple relation discussed already in Sect. 3.3 which has been explicitly been derived for the two family case. For the case that

$$t_L^{(A)} - t_L^{(B)} = t_d^{(A)} - t_d^{(B)} \tag{7.40}$$

one has the order-of-magnitude relation

$$V_{us} \sim \sqrt{\frac{m_u}{m_s}} \tag{7.41}$$

which is of a similar form as relation (3.29).

However, this type of model has various problems. Aside from the fact that the above discussion is only qualitative, the simple $U(1)_H$ model used here for illustrative purposes does not yield a sensible phenomenology. As has been discussed in [9], one needs to extend the symmetry to satisfy the phenomenological constraints from the observed masses and mixings.

Furthermore, a spontaneously broken *global* horizontal symmetry will result in massless (or at least light) scalar fields, which are not observed. This can in principle be avoided by elevating the global horizontal symmetry into a *local* one, in which case one can trade the massless modes for the longitudinal modes of massive gauge bosons. However, the masses of these gauge bosons have to be very large in order to avoid problems with flavour-changing currents.

Going beyond these qualtitative remarks is beyond the scope of this book; in particular, the ansatz discussed above is again only an effective field theory, which cannot explain the origin of the higher-dimensional operators. However, its nice feature is a qualitative explanation of the hierarchical structure of the mass matrices. Still, it is fair to say that at present there is no working model that explains the flavour structure of the Standard Model.

References

1. M. E. Peskin and T. Takeuchi, Phys. Rev. Lett. **65**, 964 (1990).
2. W. Buchmuller and D. Wyler, Nucl. Phys. B **268**, 621 (1986).
3. C. Arzt, M. B. Einhorn and J. Wudka, Nucl. Phys. B **433**, 41 (1995) [arXiv:hep-ph/9405214].
4. H. Georgi, Nucl. Phys. B **363**, 301 (1991).
5. G. D'Ambrosio, G. F. Giudice, G. Isidori and A. Strumia, Nucl. Phys. B **645**, 155 (2002) [arXiv:hep-ph/0207036].
6. T. Hansmann and T. Mannel, Phys. Rev. D **68**, 095002 (2003) [arXiv:hep-ph/0306043].
7. C. D. Froggatt and H. B. Nielsen, Nucl. Phys. B **147**, 277 (1979).
8. M. Leurer, Y. Nir and N. Seiberg, Nucl. Phys. B **398**, 319 (1993) [arXiv:hep-ph/9212278].
9. M. Leurer, Y. Nir and N. Seiberg, Nucl. Phys. B **420**, 468 (1994) [arXiv:hep-ph/9310320].

8 Prospects

8.1 Current and Future Experiments

Flavour physics is currently one of the most active fields in particle physics. This is mainly due to the fact that a lot of experimental activity has gathered an impressive amount of data.

After the discovery of the upsilon resonances, it did not take too long to use cross-section enhancement through the $\Upsilon(4S)$ to produce a significant number of bottom mesons. Two experiments opened the road to today's precision flavour physics: The ARGUS experiment at Hamburg and the CLEO experiment at Cornell. These experiments were very successful in giving us our first insights into the rich phenomenology of B meson physics. In particular, it is interesting to note that the first hint of a very large top-quark mass of far above 100 GeV came from the discovery at ARGUS (together with the UA1 experiment at CERN) of B–$\overline{B}$ oscillations. This happened at a time when the top quark was still believed to lie just above the cms-energy of the PETRA ring at Hamburg, which was at that time in the region of 30 to 40 GeV.

Both ARGUS and CLEO were successful in measuring details of semileptonic and non-leptonic decays, thereby also proving the existence of charmless decays. However, an breakthrough not originally expected came through the development of silicon vertex devices, which allowed very precise tracking at the experiments performed at the LEP collider at CERN and at the Tevatron at Fermilab. This opened the road to identifying B mesons (or more generally b quarks) in the decay products of the Z_0 or in hadronic collisions at very high energies, and so these experiments collected results for a significant number of B mesons and were able to identify all kinds of decay modes. Through this technological development the high-energy colliders could contribute to B physics significantly; for example, the lifetime values for bottom hadrons are still dominated by the collider measurements.

One key point in flavour physics could not, however, be checked by the symmetric B factories at Hamburg and Cornell, which was the measurement of the CP violation in the B system. This was the motivation for constructing asymmetric B factories with much higher luminosities, which can perform a measurement of the time-dependent CP asymmetry in B decays. It clearly marked a milestone in flavour physics when the BaBar experiment at the

SLAC B factory and the Belle experiment at the KEK B factory measured the time-dependent CP asymmetry in $B \to J/\Psi K_s$ in 2001. In the meantime, the measurement of the CKM angle β has become a precision measurement with an uncertainty of about 5%. Unfortunately, the measured CP asymmetry is in accord with the prediction of the CKM picture.

Currently, the B factories are producing an enormous amount of data. The analysis of these data will also boost the theoretical developments aimed at obtaining a better understanding of non-leptonic decays. This is obviously of vital importance in overconstraining the unitarity triangle by a measurement of the two other CKM angles α and γ. Still it will be a theoretical challenge to assess the uncertainties for extracting α and γ from the observed CP asymmetries.

Similarly, the measurement of rare decay modes is important for a further understanding of flavour. In particular, FCNC modes such as those based on the quark transition $b \to s\gamma$ will give us deeper insight into the GIM mechanism and possibly open a window onto new physics.

Clearly, the B factories alone will not allow a complete check of the flavour sector. It is important to include data on B_s decays in the analysis. However, the B factories are running below the B_s threshold and thus second-generation B physics experiments are needed.

These experiments are on their way. On one hand, the experiments at the Tevatron will provide a significant sample of B_s mesons and the discovery of B_s–$\overline{B}_s$ oscillations is around the corner and is just a matter of sufficient integrated luminosity. A measurement of these oscillations will give us another important constraint on the unitarity triangle, which cannot be obtained from the B factories. On the other hand, there are plans to have dedicated experiments at the LHC (LHC-b) and at the Tevatron (B-TeV), which are designed to give us precise information on B_s decays also.

To get a full picture, it will be necessary to measure various B_s decays, and, in particular, the CP asymmetry in these decays. The mode $B_s \to J/\Psi\phi$ plays a similar role in the B_s system to the mode $B \to J/\Psi K_s$ in the B system: it measures the phase of the mixing, which is predicted to be very small. Clearly a large time-dependent CP asymmetry would be a clear signal of new physics. Similarly, many strategies to extract γ using flavour symmetries need information on B_s decays.

In addition, there have been initial discussions about increasing the luminosity of the SLAC and KEK B factories by a factor of about 100 to allow the measurement of B decays with very small branching ratios. Even if the number of B mesons produced is comparable to the numbers at LHC-b or B-TeV, the cleaner environment at an e^+e^- B factory may be an advantage.

8.2 Theoretical Perspectives

With this enormous amount of experimental information, we are entering the era of precision flavour physics. The theoretical methods based on effective field theories are in a mature state, and many effective-theory approaches have been studied extensively.

As far as the effective Hamiltonian for weak interactions is concerned, the full renormalization group result is known to next-to-leading order. In most cases this is sufficient, since the limitations arise from the matrix elements of the operators. However, in a few cases, such as the FCNC process $B \to X_s \gamma$, it turns out that the theoretical prediction obtained from the next-to-leading-order calculation is not sufficient. Thus we need to go beyond the next-to-leading order which is a technical challenge, since it requires a four-loop calculation of an anomalous dimension, a three-loop matching calculation and the calculation of the matrix element at the two-loop level.

Some progress has been achieved by applying effective-field-theory methods to QCD, exploiting the fact that the mass of the b quark is large compared with the scale parameter Λ_{QCD}. The classical applications known as "heavy-quark effective theory" and "heavy-quark expansion" can be applied to both exclusive and inclusive processes. The applications in which the light degrees of freedom are treated as soft are in a mature state and have resulted in a precise determination of V_{cb}, which is by now the second-best-known CKM matrix element. With more data, a clean determination of V_{ub} will become possible, since the theoretical tools exist.

More recently, ideas have been discussed for using an effective-field-theory picture for a situation in which the light degrees of freedom have large momenta and result in collimated, energetic jets of light hadrons or even, in the exclusive case, in a single, energetic light hadron. These developments are known under the name of "soft collinear effective theory" (SCET) and are currently under investigation.

The main impact of these new ideas is that SCET opens the way to a QCD-based calculation of exclusive non-leptonic processes. The key ingredients are factorization theorems that allow us to relate different processes, thereby eliminating non-perturbative quantities. However, currently the data indicatie that the leading terms in the SCET expansion might receive substantial subleading corrections and it is currently unclear how reliable this method is.

Combining the theoretical methods with the current data has already provided a significant test of the CKM picture. The current constraints in the ρ–η plane are shown in Fig. 8.1. The plot shows the remarkable consistency of the current data, which do not show any significant deviation from the CKM picture of the Standard Model. As of today, there are only a few "hints" which do not have any statistical significance (see [2]).

In order to discuss effects beyond the Standard Model, one can interpret the Standard Model itself as the renormalizable piece of an effective field

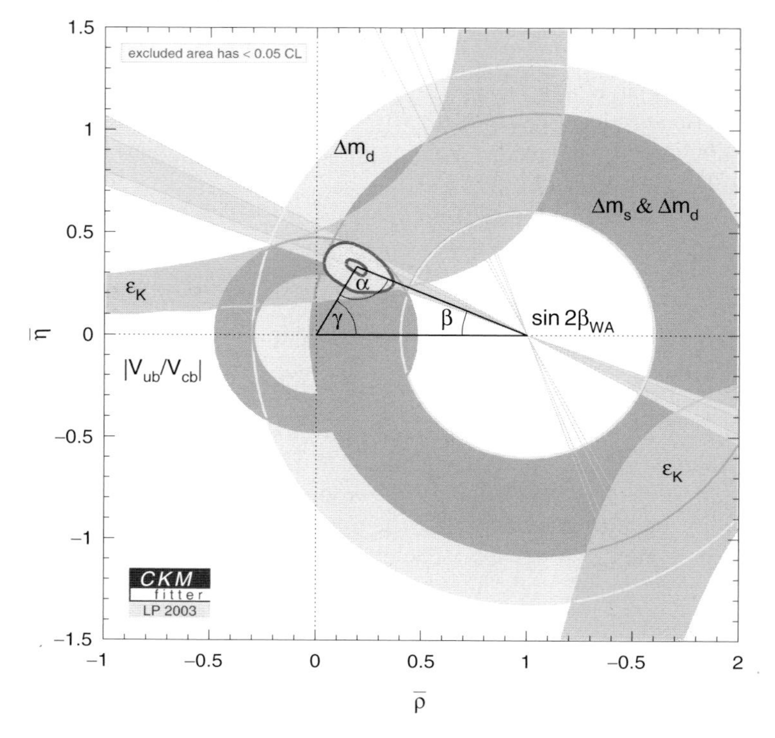

Fig. 8.1. Current constraints in the ρ-η. The figure is taken from [1]

theory. From this point of view any physics beyond the Standard Model is parametrized by operators of higher dimension. While this idea has been applied with great success to the gauge sector, it is not well suited to the flavour sector, since the number of unknown parameters is too large to make this approach useful.

A restriction on the number of parameters requires to assume some model framework for the new physics, of which we do not have any indication from precision measurements. However, as far as flavour is concerned, it is not easy to invent a mechanism which explains the hierarchy of masses and mixing angles. It is fair to say that no plausible model of flavour exists.

Effective-field-theory methods have developed into a standard tool in particle physics, and a lot of the theoretical progress has originated from a clever use of these methods. On the other hand, a stringent test of the flavour sector of the Standard Model requires further work; as far as experiments are concerned, the experimental uncertainties will become really small at the end of the B-factory era, at least for a large variety of observables. It will be a challenge for the theoretical side to match this precision; this, however, will be vital for exploiting the full information contained in the experimental data that will be available ten years from now.

References

1. J. Charles et al. [CKM Fitter working group], http://ckmfitter.in2p3.fr/
2. Y. Grossman, plenary talk at the 21st International Symposium on Lepton and Photon Interactions at High Energies (LP 03), Batavia, Illinois, 11–16 Aug. 2003, [arXiv:hep-ph/0310229].

Index

Springer Tracts in Modern Physics

183 **Transverse Patterns in Nonlinear Optical Resonators**
By K. Staliūnas, V. J. Sánchez-Morcillo 2003. 132 figs., XII, 226 pages

184 **Statistical Physics and Economics**
Concepts, Tools and Applications
By M. Schulz 2003. 54 figs., XII, 244 pages

185 **Electronic Defect States in Alkali Halides**
Effects of Interaction with Molecular Ions
By V. Dierolf 2003. 80 figs., XII, 196 pages

186 **Electron-Beam Interactions with Solids**
Application of the Monte Carlo Method to Electron Scattering Problems
By M. Dapor 2003. 27 figs., X, 110 pages

187 **High-Field Transport in Semiconductor Superlattices**
By K. Leo 2003. 164 figs.,XIV, 240 pages

188 **Transverse Pattern Formation in Photorefractive Optics**
By C. Denz, M. Schwab, and C. Weilnau 2003. 143 figs., XVIII, 331 pages

189 **Spatio-Temporal Dynamics and Quantum Fluctuations in Semiconductor Lasers**
By O. Hess, E. Gehrig 2003. 91 figs., XIV, 232 pages

190 **Neutrino Mass**
Edited by G. Altarelli, K. Winter 2003. 118 figs., XII, 248 pages

191 **Spin-orbit Coupling Effects in Two-dimensional Electron and Hole Systems**
By R. Winkler 2003. 66 figs., XII, 224 pages

192 **Electronic Quantum Transport in Mesoscopic Semiconductor Structures**
By T. Ihn 2003. 90 figs., XII, 280 pages

193 **Spinning Particles – Semiclassics and Spectral Statistics**
By S. Keppeler 2003. 15 figs., X, 190 pages

194 **Light Emitting Silicon for Microphotonics**
By S. Ossicini, L. Pavesi, and F. Priolo 2003. 206 figs., XII, 284 pages

195 **Uncovering *CP* Violation**
Experimental Clarification in the Neutral K Meson and B Meson Systems
By K. Kleinknecht 2003. 67 figs., XII, 144 pages

196 **Ising-type Antiferromagnets**
Model Systems in Statistical Physics and in the Magnetism of Exchange Bias
By C. Binek 2003. 52 figs., X, 120 pages

197 **Electroweak Processes in External Electromagnetic Fields**
By A. Kuznetsov and N. Mikheev 2003. 24 figs., XII, 136 pages

198 **Electroweak Symmetry Breaking**
The Bottom-Up Approach
By W. Kilian 2003. 25 figs., X, 128 pages

199 **X-Ray Diffuse Scattering from Self-Organized Mesoscopic Semiconductor Structures**
By M. Schmidbauer 2003. 102 figs., X, 204 pages

200 **Compton Scattering**
Investigating the Structure of the Nucleon with Real Photons
By F. Wissmann 2003. 68 figs., VIII, 142 pages

201 **Heavy Quark Effective Theory**
By A. Grozin 2004. 72 figs., X, 213 pages

202 **Theory of Unconventional Superconductors**
By D. Manske 2004. 84 figs., XII, 228 pages

203 **Effective Field Theories in Flavour Physics**
By T. Mannel 2004. 29 figs., VIII, 175 pages

Printing: Saladruck, Berlin
Binding: Stein+Lehmann, Berlin